Lecture Notes in Mathematics

A collection of informal reports and seminars
Edited by A. Dold, Heidelberg and B. Eckmann, Zürich

T0222250

146

Allen Altman

University of California, La Jolla, CA/USA

Steven Kleiman

M.I.T., Dept. of Mathematics, Cambridge, MA/USA

Introduction to Grothendieck Duality Theory

Springer-Verlag
Berlin · Heidelberg · New York 1970

PREFACE

These notes grew out of a Columbia seminar on Grothendieck's Bourbaki talk [6] on duality and his SGA talks [9] on flat, étale, and smooth morphisms. They are intended as a second course in algebraic geometry and assume only a general familiarity with schemes including Serre's theorems on the cohomology of projective space. The central result follows:

<u>Theorem</u>. Let k be a field and X a projective k-scheme of pure dimension r. Then there exist uniquely a coherent O_X-Module ω_X and a "residue" map $\eta_X: H^r(X,\omega_X) \longrightarrow k$ such that, for any coherent O_X-Module F and integer p, there exists a canonical pairing

$$H^p(X,F) \times \mathrm{Ext}^{r-p}_{O_X}(F,\omega_X) \longrightarrow H^r(X,\omega_X) \xrightarrow[=]{\eta_X} k$$

which is always nonsingular for $p = r$ and is nonsingular for all p if and only if X is Cohen-Macaulay. Furthermore, if X is a closed subscheme of $P = \mathbb{P}^n_k$, then $\omega_X = \mathrm{Ext}^{n-k}_{O_P}(O_X, O_P(-n-1))$; if X is smooth over k, then $\omega_X = \Omega^r_{X/k}$; and if X is a smooth curve, then η_X is defined by the classical residue symbol.

The material divides naturally into four parts. The first part, (Chapter I), presupposing the others, discusses ω_X. The second, (Chapters II, III, IV), first develops preliminaries of commutative and homological algebra; it then establishes the duality theorems. The third part, (Chapters V, VI, VII), studies smooth morphisms aiming for general familiarity. (Lacking notably, however, is a proof of Zariski's Main Theorem and application to the branch locus of covers of normal schemes). Finally, the last part, (Chapter VIII), treating curves, gives the traditional construction of ω_X and proof of duality, and, using Tate's elegant approach [13], it proves η_X arises from residues.

Allen Altman

Steven Kleiman

New York, 1968

CONTENTS

Chapter I - Study of ω_X

1. Main Duality Results

<u>(1.1) Yoneda pairing (IV,1)</u>. - Let X be a ringed space and F,ω two O_X-Modules. Then there exists a ∂-functorial pairing

$$H^p(X,F) \times \mathrm{Ext}^{r-p}_{O_X}(F,\omega) \longrightarrow H^r(X,\omega)$$

for all integers r,p. Furthermore, if F is locally free of finite rank, the pairing becomes:

$$H^p(X,F) \times H^{r-p}(X,\omega \otimes F^\vee) \longrightarrow H^r(X,\omega).$$

<u>(1.2) Serre duality (IV,4)</u>. - Let k be a field, $P = \mathbb{P}^n_k$, projective n-space over k, F a coherent O_P-Module and $\omega_P = O_P(-n-1)$. If $\eta_P : H^n(P,\omega_P) \longrightarrow k$ is a fixed isomorphism, then the Yoneda pairing, composed with η_P, defines a ∂-functorial pairing which is nonsingular, or, equivalently, the corresponding map

$$\mathrm{Ext}^{n-p}_{O_P}(F,\omega_P) \longrightarrow H^p(P,F)^*$$

is an isomorphism of ∂-functors.

<u>(1.3) Grothendieck duality (IV,5)</u>. - Let k be a field, $P = \mathbb{P}^n_k$, projective n-space over k, and X a closed subscheme of P of pure dimension r. Let F be a coherent O_X-Module, $\omega_P = O_P(-n-1)$ and $\omega_X = \underline{\mathrm{Ext}}^{n-r}_{O_P}(O_X,\omega_P)$. Then an isomorphism $\eta_P : H^n(P,\omega_P) \longrightarrow k$ defines a map $\eta_X : H^r(X,\omega_X) \longrightarrow k$, which, composed with the Yoneda pairing, yields a pairing

$$H^p(X,F) \times \mathrm{Ext}^{r-p}_{O_X}(F,\omega_X) \longrightarrow k.$$

For $p = r$, this pairing is always nonsingular. For $r - s \leq p \leq r$, it is nonsingular if and only if $\underline{\mathrm{Ext}}^{n-p}_{O_P}(O_X,\omega_P) = 0$. In particular, it is nonsingular for all p if and only if X is Cohen-Macaulay (e.g., X regular or, more generally, locally a complete intersection in P).

Furthermore, (I,4.6), if X is smooth over k, then $\omega_X = \Omega^r_{X/k}$, and (VIII,4.4), if X is a smooth curve, then η_X is defined by the classical residue symbol.

2. Further discussion of ω_X

<u>Proposition (2.1)</u>. - Under the conditions of (1.3), the pair (η_X,ω_X) is a character of X, uniquely determined up to unique isomorphism.

<u>Proof</u>. The assertion results formally from the following lemma.

<u>Lemma (2.2)</u>. - Under the conditions of (1.3), for any map $\varphi : H^r(X,F) \longrightarrow k$, there exists a unique map $f : F \longrightarrow \omega_X$ making the following diagram commute:

$$\begin{array}{ccc} H^r(X,F) & \xrightarrow{H^r(X,f)} & H^r(X,\omega_X) \\ & \searrow^{\varphi} \quad \swarrow_{\eta_X} & \\ & k & \end{array}$$

<u>Proof</u>. The assertion results immediately from (1.3).

<u>Proposition (2.3)</u>. - Let P be a regular k-scheme of pure dimension n and Y (resp. X), a closed subscheme of P (resp. Y) of

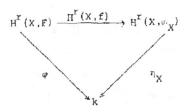

pure dimension s (resp. r). Let ω_P be an invertible sheaf on P, $\omega_X = \underline{Ext}^{n-r}_{O_P}(O_X,\omega_P)$ and $\omega_Y = \underline{Ext}^{n-s}_{O_P}(O_Y,\omega_P)$. If Y is Cohen-Macaulay, then $\omega_X = \underline{Ext}^{s-r}_{O_Y}(O_X,\omega_Y)$.

Proof. By (III,5.22) and (IV,5.1), $\underline{Ext}^q_{O_P}(O_Y,\omega_P) = 0$ for $q \neq n-s$; so, the spectral sequence (IV,2.9.2)

$$\underline{Ext}^p_{O_Y}(O_X,\underline{Ext}^q_{O_P}(O_Y,\omega_P)) \Rightarrow \underline{Ext}^{p+q}_{O_P}(O_X,\omega_P)$$

degenerates and yields a canonical isomorphism

$$\underline{Ext}^{s-r}_{O_Y}(O_X,\omega_Y) \xrightarrow{\sim} \underline{Ext}^{n-r}_{O_P}(O_X,\omega_P).$$

Proposition (2.4). - Let X be a scheme and D an effective divisor, considered as a closed subscheme of X. Let ω_X be an O_X-Module and $\omega_D = \underline{Ext}^1_{O_X}(O_D,\omega_X)$. Then there exists a natural isomorphism

$$O_D \otimes_{O_X} (O_X(D) \otimes_{O_X} \omega_X) \xrightarrow{\sim} \omega_D.$$

In particular, if ω_X is locally free, then ω_D is locally free.

Proof. The exact sequence (VII,3.6)

$$0 \longrightarrow O_X(-D) \longrightarrow O_X \longrightarrow O_D \longrightarrow 0$$

yields the diagram

$$\begin{array}{ccccccc}
\omega_X & \longrightarrow & O_X(D) \otimes_{O_X} \omega_X & \longrightarrow & O_D \otimes_{O_X} (O_X(D) \otimes_{O_X} \omega_X) & \longrightarrow & 0 \\
\downarrow \wr & & \downarrow \wr & & \vdots \downarrow & & \\
\underline{Hom}_{O_X}(O_X,\omega_X) & \longrightarrow & \underline{Hom}_{O_X}(O_X(-D),\omega_X) & \longrightarrow & \underline{Ext}^1_{O_X}(O_D,\omega_X) & \longrightarrow & 0
\end{array}$$

whence, the assertion.

Remark (2.5). - Under the conditions of (1.3), if X is smooth, $O_X(D)\otimes_{O_X}\omega_X$ may be interpreted as the sheaf of differentials on X with poles only along D (the order bounded by D). The homomorphism $O_X(D)\otimes_{O_X}\omega_X \longrightarrow \omega_D$ is often called the Poincaré residue map.

Corollary (2.6). - Let P be a scheme, X a closed subscheme and ω_P a locally free O_P-Module. If X is regularly immersed in P of pure codimension $n-r$, then $\omega_X = \underline{\mathrm{Ext}}^{n-r}_{O_P}(O_X,\omega_P)$ is locally free.

Proof. Since the assertion is local, we may assume X is "cut out" by a regular sequence of elements $f_1,\dots,f_{n-r}\in\Gamma(P,O_P)$. Let D_i be the closed subscheme of D_{i-1} "cut out" by f_i. Then D_i is a divisor on D_{i-1} and the assertion follows from (2.4)

Proposition (2.7). - Let P be a regular scheme, X a closed subscheme of pure codimension $n-r$, ω_P an invertible sheaf on P and $\omega_X = \underline{\mathrm{Ext}}^{n-r}_{O_P}(O_X,\omega_P)$. Suppose X is generically reduced. Then there exists an open dense subset U of X such that $\omega_X|U$ is locally free of rank 1.

Proof. If J is the ideal defining X, then, at any generic point x of X, $J_x = m_x$. So, since $O_{P,x}$ is regular of dimension $n-r$, J is generated by $n-r$ elements on an open set U about x. The assertion now follows from (III,4.5 and 4.12) and (2.4).

Proposition (2.8). - Under the conditions of (1.3), if X is reduced, then ω_X is torsion free of rank 1.

Proof. Let K_X be the sheaf of rational functions on X and define F by $0 \longrightarrow F \xrightarrow{f} \omega_X \longrightarrow \omega_X \otimes_{O_X} K_X$. By (2.7), there exists an open dense subset U on which ω_X is invertible. Then $\mathrm{Supp}\,(F) \subset X - U$, so $\dim\,(\mathrm{Supp}\,(F)) < r$. Therefore, $H^r(X,F) = 0$; so, by (1.3), $\mathrm{Hom}_{O_X}(F,\omega_X) = 0$. Hence, $f = 0$ and $F = 0$.

Lemma (2.9). - Under the conditions of (1.3), let $X_1,\ldots,X_p$ be the irreducible components of X and x_i the generic point of X_i. Then the canonical map

$$\mathrm{Hom}_{O_X}(F,\omega_X) \longrightarrow \Pi\ \mathrm{Hom}_{O_{X_i,x_i}}(F_{x_i},\omega_{X,x_i})$$

is injective.

Proof. Let $f : F \longrightarrow \omega_X$ be a homomorphism such that the maps $f_{x_i} : F_{x_i} \longrightarrow \omega_{X,x_i}$ are all zero, and let $G = \mathrm{Im}(f)$. Since $x_i \notin \mathrm{Supp}(G)$, $\dim(\mathrm{Supp}(G)) < r$. Hence, $H^r(X,G) = 0$; so by (1.3), $\mathrm{Hom}_{O_X}(G,\omega_X) = 0$. Since $G \hookrightarrow \omega_X$, it follows that $G = 0$; whence, the assertion.

Proposition (2.10). - Under the conditions of (1.3), suppose that X is integral and that k is algebraically closed in the function field K of X Then η_X is an isomorphism.

Proof. If x is the generic point of X, then $K = O_{X,x}$. Hence by (2.9), and (2.7), the canonical map

$$A = \mathrm{Hom}_{O_X}(\omega_X,\omega_X) \longrightarrow \mathrm{Hom}_K(K,K) = K$$

is injective. However, by (IV,3.2), A is a finite dimensional k-algebra. Thus $A = k$ and, by (1.3), $H^r(X,\omega_X) = k$; whence, the assertion.

3. Differentials on Projective Space.

Let S be a scheme, X an S-scheme and $0 \longrightarrow E' \longrightarrow E \xrightarrow{u} E'' \longrightarrow 0$ a locally split, exact sequence of quasi-coherent O_X-Modules. Let $Z = \mathbb{V}(E)$ $(=\mathrm{Spec}(S(E)))$, $Y = \mathbb{V}(E'')$ and $J = \ker(S(u))$. Then $\tilde{J}$ is the O_Z-ideal defining the closed immersion $Y \hookrightarrow Z$. The map $\delta : \mathcal{J}/\mathcal{J}^2 \longrightarrow \Omega^1_{Z/S} \otimes_{O_Z} O_Y$, defined by $d_{Z/S}$, induces a map $J/J^2 \longrightarrow f_*(\Omega^1_{Z/S}) \otimes_{S(E)} S(E'')$, where $f : Z \longrightarrow X$ is the structure map; hence, a map $\alpha'' : E' \longrightarrow f_*(\Omega^1_{Z/S}) \otimes_{S(E)} S(E'')$.

Assume $E = O_X \otimes_{O_S} F$ where F is a quasi-coherent O_S-Module and let $V = \mathbb{V}(F)$. Then $Z = X \times_S V$; so, by (VI,1.12), $\Omega^1_{Z/S} = (\Omega^1_{X/S} \otimes_{O_S} O_V) \oplus (O_X \otimes_{O_S} \Omega^1_{V/S})$ and $d_{Z/S} = (d_{X/S} \otimes id_V) + (id_X \otimes d_{V/S})$. The map α'', followed by projection on the first factor, yields a map $E' \longrightarrow f_*(f^*\Omega^1_{X/S}) \otimes_{S(E)} S(E'')$. If $\Omega^1_{X/S}$ is locally free of finite rank, the canonical map $\Omega^1_{X/S} \otimes_{O_X} S(E) \longrightarrow f_* f^* \Omega^1_{X/S}$ is an isomorphism; so, the above map becomes $\alpha' : E' \longrightarrow \Omega^1_{X/S} \otimes_{O_X} S(E) \otimes_{S(E)} S(E'')$

To compute α' locally, assume S and X are affine and let $e' = \Sigma a_i \otimes_{O_S} t_i \in \Gamma(X,E')$ where $a_i \in \Gamma(X,O_X)$ and $t_i \in \Gamma(S,F)$. Then $\alpha'(e') = \Sigma da_i \otimes_{O_X} (1 \otimes_{O_S} t_i) \otimes_{S(E)} 1 = \Sigma\, da_i \otimes_{O_X} u(1 \otimes t_i)$ where $u : E \longrightarrow E''$; so, $\alpha'(e')$ is a global section of $\Omega^1_{X/S} \otimes_{O_X} E''$. Thus α' induces a map

$$\alpha : E' \longrightarrow \Omega^1_{X/S} \otimes_{O_X} E'',$$

called the <u>second fundamental form of</u> E' <u>in</u> E.

<u>Theorem (3.1)</u>. - Let S be a scheme, F a locally free O_S-Module of finite rank and $P = \mathbb{P}(F)$. Let $p : P \longrightarrow S$ be the structure map and $u : p^*F \longrightarrow O_P(1)$, the canonical surjection. Then the second fundamental form of $\mathrm{Ker}(u)$ in p^*F gives rise to an exact sequence

$$0 \longrightarrow \Omega^1_{P/S}(1) \longrightarrow p^*F \xrightarrow{u} O_P(1) \longrightarrow 0.$$

Furthermore, if F is free of rank $n+1$, then $\Omega^n_{P/S} = \Lambda^n \Omega^1_{P/S}$ is canonically isomorphic to $O_P(-n-1)$.

<u>Proof</u>. Let $E' = \ker(u)$; we prove that $\alpha : E' \longrightarrow \Omega^1_{P/S}(1)$ is an isomorphism. Note that by (VII,5.1), $\Omega^1_{P/S}$ is locally free of finite rank; hence, α is defined. We may work locally and so assume S is affine with ring A and $P = \mathrm{Proj}(A[T_0,\ldots,T_n])$ where the T_i are indeterminates. Consider the open affine $U = D_+(T_j)$ of P whose ring is $B = A\left[\frac{T_0}{T_j},\ldots,\frac{T_n}{T_j}\right]$. If $F = O_S e_0 \oplus \ldots \oplus O_S e_n$, then $u(e_i) = \frac{T_i}{T_j} T_j \in BT_j = \Gamma(U, O_P(1))$. Hence, $\Gamma(U,E')$ is the free B-module with basis $e_i' = \frac{T_i}{T_j} e_j - e_i$, $i \neq j$. The form α is given by $\alpha(e_i') = d\left(\frac{T_i}{T_j}\right) \otimes T_j \in \Gamma(U, \Omega^1_{P/S}(1))$; since, by (VI,1.4), the elements $d\left(\frac{T_i}{T_j}\right) \otimes T_j \, (i \neq j)$ form a basis of $\Gamma(U, \Omega^1_{P/S}(1))$, α is an isomorphism. The last assertion now follows easily from (VII,3.12).

4. The Fundamental Local Isomorphism

Definition (4.1). - Let A be a ring and $x_1,\dots,x_r \in A$. The Koszul complex $K_*(\underline{x})$ determined by $(\underline{x}) = (x_1,\dots,x_r)$ is defined as follows: $K_p(\underline{x}) = \wedge^p(\overset{r}{\underset{i=1}{\oplus}} Ae_i)$ for $0 \leq p \leq r$ and $K_p(\underline{x}) = 0$ otherwise. The boundary map $d_p: K_p(\underline{x}) \longrightarrow K_{p-1}(\underline{x})$ is defined by

$$d_p(e_{i_1}\wedge\dots\wedge e_{i_p}) = \Sigma(-1)^j x_{i_j} e_{i_1}\wedge\dots\wedge\hat{e}_{i_j}\wedge\dots\wedge e_{i_p}.$$

Lemma (4.2). - Let A be a ring, $(x_1,\dots,x_r)$ an A-regular sequence, $I = x_1A+\dots+x_rA$, and M an A-module. Then $K_*(\underline{x};M) = K_*(\underline{x})\otimes_A M$ is a resolution of M/IM.

Proof. Note that $K_*(\underline{x};M)$ is the (single) complex associated to the double complex $K^{p,q} = K_p((x_1,\dots,x_{r-1});M)\otimes K_q(x_r)$. Further, we may assume by induction on r that ${}_{II}E_1^{p,q} = H_I^p(K^{*,q}) = 0$ for $(p,q) \neq (0,0)$ or $(0,1)$ and ${}_{II}E_1^{0,q} = M/I'M$ for $q = 0,1$ where I' is the ideal generated by $x_1,\dots,x_{r-1}$. By assumption, $x_r: M/I'M \longrightarrow M/I'M$ is injective; so ${}_{II}E_2^{p,q} = 0$ for $(p,q) \neq (0,0)$ and ${}_{II}E_2^{0,0} = M/IM$. Since ${}_{II}E^{p,q} \Longrightarrow H^{p+q}(K_*(\underline{x};M))$, $K_*(\underline{x};M)$ is a resolution of M/IM.

Lemma (4.3). - Let A be a ring, M an A-module and $x_1,\dots x_r \in A$. Set $K^*(\underline{x};M) = \mathrm{Hom}_A(K_*(\underline{x});M)$ and $H^p(\underline{x};M) = H^p(K^*(\underline{x};M))$ and define $\varphi'_{\underline{x}} : K^r(\underline{x};M) \longrightarrow M$ by $\varphi'_{\underline{x}}(a) = a(e_1\wedge\dots\wedge e_r)$. Then $\varphi'_{\underline{x}}$ induces an isomorphism

$$\varphi_{\underline{x}} : H^r(\underline{x};M) \xrightarrow{\sim} M/IM$$

where I is the ideal generated by $x_1,\dots,x_r$.

Proof. Note that $\varphi'_{\underline{x}}(d(b)) \in IM$ for each $b \in K^{r-1}(\underline{x};M)$; thus,

$\varphi'_{\underline{x}}$ induces the required map $\varphi_{\underline{x}}$. It is clearly surjective. Suppose $\varphi'_{\underline{x}}(a) = 0$. Then $a(e_1\wedge\ldots\wedge e_r) = \Sigma x_j y_j$ for suitable $y_j \in M$. Define $b : \wedge^{r-1}(A^r) \longrightarrow M$ by $b(e_1\wedge\ldots\wedge\hat{e}_j\wedge\ldots\wedge e_r) = (-1)^j y_j$. Clearly, $d(b) = a$ and, hence, $\varphi_{\underline{x}}$ is injective.

Lemma (4.4). - Let A be a ring, M an A-module and I an ideal of A. Let $(x_1,\ldots,x_r)$ and $(y_1,\ldots,y_r)$ be two A-regular sequences which generate I and let $y = \Sigma c_{ij} x_j$ where $c_{ij} \in A$. Then there exists a commutative diagram

$$\begin{array}{ccc} & & M/IM \\ & \nearrow^{\varphi_{\underline{x}}} & \\ \mathrm{Ext}^r_A(A/I,M) & & \Big\downarrow \det(c_{ij}) \\ & \searrow_{\varphi_{\underline{y}}} & \\ & & M/IM \end{array}$$

Proof. Since $(\underline{x})$ and $(\underline{y})$ are A-regular, $\mathrm{Ext}^r_A(A/I,M) = H^r(\underline{x};M) = H^r(\underline{y};M)$ by (4.3). Furthermore, $\wedge c : K_*(\underline{y}) \longrightarrow K_*(\underline{x})$ is a ∂-isomorphism. Since $\wedge^r c = \det(c_{ij})$, the commutativity results from the definitions.

Theorem (4.5). - Let P be a scheme, X a closed subscheme, J its sheaf of ideals and F a quasi-coherent O_X-Module. Suppose X is regularly immersed in P. Then there exists a natural isomorphism

$$\varphi : \underline{\mathrm{Ext}}^r_{O_P}(O_X,F) \xrightarrow{\sim} \underline{\mathrm{Hom}}_{O_X}(\wedge^r(J/J^2),F/JF)$$

where $r = \mathrm{codim}(X,P)$.

Proof. Let U be an affine open set of P on which J is a regular ideal; let A be the ring of U, $M = \Gamma(U,F)$ and $I = \Gamma(U,J)$.

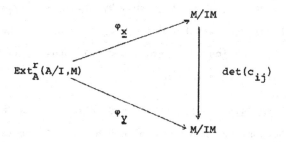

Then I is generated by an A-regular sequence $(x_1,\ldots,x_r)$ and I/I^2 is free of rank r over A/I by (III,3.4); hence, the exterior product $x_1'\wedge\ldots\wedge x_r'$ of the residue classes x_i' generates $\wedge^r(I/I^2)$ and we may define

$$\varphi : \mathrm{Ext}^r_A(A/I,M) \longrightarrow \mathrm{Hom}_{A/I}(\wedge^r(I/I^2),M/IM) \text{ by}$$

$$\varphi(a)(x_1'\wedge\ldots\wedge x_r') = \varphi_{\underline{x}}(a).$$

If $(y_1,\ldots,y_r)$ is another A-regular sequence that generates I, then there exist $c_{ij}\in A$ such that $y_i = \Sigma c_{ij}x_j$. Then $y_1\wedge\ldots\wedge y_r = \det(c_{ij})x_1\wedge\ldots\wedge x_r$ and, by (4.4), $\varphi(a)(y_1\wedge\ldots\wedge y_r) = $
$= \det(c_{ij})\varphi(a)(x_1\wedge\ldots\wedge x_r) = \det(c_{ij})\varphi_{\underline{x}}(a) = \varphi_{\underline{y}}(a)$. Hence, φ is independent of choice of generators of I and, by (IV,3.2), φ defines a global isomorphism.

Theorem (4.6). - Let P be an S-scheme and X a closed subscheme. Suppose X and P are smooth over S of relative dimensions n and r. Then

$$\Omega^r_{X/S} = \underline{\mathrm{Ext}}^{n-r}_{O_P}(O_X,\Omega^n_{P/S})$$

In particular, if $P = \mathbb{P}^n_S$ and $\omega_X = \underline{\mathrm{Ext}}^{n-r}_{O_P}(O_X,O_P(-n-1))$, then

$$\omega_X = \Omega^r_{X/S}.$$

Proof. By (VII,5.13), X is regularly immersed in P. Hence, by (4.5) and (IV,3.4), $\underline{\mathrm{Ext}}^{n-r}_{O_P}(O_X,\Omega^n_{P/S}) = \underline{\mathrm{Hom}}_{O_X}(\wedge^{n-r}(J/J^2),\Omega^n_{P/S}\otimes_{O_P}O_X) =$
$= (\wedge^{n-r}(J/J^2))^\vee\otimes_{O_P}\Omega^n_{P/S}$ where J is the sheaf of ideals of X in P. Now, by (VII,5.8), the sequence $0\rightarrow J/J^2\rightarrow \Omega^1_{P/S}\otimes_{O_P}O_X\rightarrow \Omega^1_{X/S}\rightarrow 0$ is exact. Therefore, by (VII,3.12), $\Omega^r_{X/S} = (\wedge^{n-r}(J/J^2))^\vee\otimes_{O_P}\Omega^n_{P/S}$; whence the first assertion. The second now results from assertion (3.1).

Chapter II - Completions, Primary Decomposition and Length

1. Completions

<u>Definition (1.1)</u>. - Let A be a ring. A family of ideals (A_n), $n \in \mathbb{N}$, is said to form a (descending) <u>filtration</u> of A, if $A_o = A$, $A_{n+1} \subset A_n$ and $A_n A_m \subset A_{n+m}$. Let M be an A-module. A family of submodules (M_n) is said to form a (compatible) <u>filtration</u> if $M_o = M$, $M_{n+1} \subset M_n$ and $A_m M_n \subset M_{m+n}$. The filtration (M_n) is said to be <u>separated</u> if $\cap M_n = 0$. Let q be an ideal of A. The <u>q-adic filtration</u> of A is defined by $A_n = q^n$; the <u>q-adic filtration</u> of M is defined by $M_n = q^n M$.

<u>Remark (1.2)</u>. - If A is a filtered ring, the sets A_n form a system of neighborhoods of 0 for a topology on A which is compatible with the ring structure of A. Similarly, if M is a compatibly filtered A-module, the sets M_n form a system of neighborhoods of 0 for a topology on M, which is compatible with the topology on A.

<u>Definition (1.3)</u>. - A ring A is said to be <u>graded</u> if there exists a family of subgroups (A_n) such that $A = \oplus A_n$ and $A_m A_n \subset A_{m+n}$ An A-module M is said to be (compatibly) <u>graded</u> if there exists a family of subgroups (M_n) such that $M = \oplus M_n$ and $A_m M_n \subset M_{m+n}$.

<u>Remark (1.4)</u>. - Let A be a filtered ring and M a compatibly filtered A-module. Let $gr^n(A) = A_n/A_{n+1}$, and $gr^n(M) = M_n/M_{n+1}$. Then $gr^*(A) = \oplus gr^n(A)$ is called the <u>associated graded ring</u> and $gr^*(M) = \oplus gr^n(M)$ the <u>associated graded</u> $gr^*(A)$-module. If A and M

are filtered by the q-adic filtration, we also write $gr_q^*(A)$ for $gr^*(A)$ and $gr_q^*(M)$ for $gr^*(M)$.

Lemma (1.5). - Let A be a filtered ring and $u : M \longrightarrow N$ a homomorphism of filtered A-modules, $(u(M_r) \subset N_r)$. Suppose $\cap M_r = 0$. If $gr^*(u)$ is injective, then u is injective.

Proof. Since $gr^*(u)$ is injective for each r, $M_r \cap u^{-1}(N_{r+1}) \subset M_{r+1}$. It follows by induction that $M_{r-k} \cap u^{-1}(N_{r+1}) \subset M_{r+1}$ for each r and each $k \geq 0$; in particular, for $k = r$, $u^{-1}(N_{r+1}) \subset M_{r+1}$. Therefore, $u^{-1}(0) \subset \cap u^{-1}(N_r) \subset \cap M_r = 0$.

Definition (1.6). - Let A be a ring. A collection of A-modules $\{M_i\}$ and A-homomorphisms $f_i^{i+1} : M_{i+1} \longrightarrow M_i$, $i \geq 0$, is said to be a projective system of A-modules indexed by $\mathbb{N}$. The projective (or inverse) limit of $\{M_i, f_i^{i+1}\}$, denoted $\varprojlim M_i$, is an A-module M together with maps $f_i : M \longrightarrow M_i$ such that $f_i^{i+1} \circ f_{i+1} = f_i$ for all i satisfying the following universal property:

> If M' is an A-module together with maps $g_i : M' \rightarrow M_i$ such that $f_i^{i+1} \circ g_{i+1} = g_i$ for all i, then there exists a unique map $g : M' \longrightarrow M$ such that $g_i = f_i \circ g$.

Proposition (1.7). - (i) Let $\{M_i, f_i^{i+1}\}$ be a projective system of A-modules indexed by $\mathbb{N}$. Then the projective limit exists.

(ii) Let N be a filtered A-module with filtration (N_n). Then the projective limit $\varprojlim N/N_n$ is the topological, separated completion $\hat{N}$, (namely, the set of Cauchy sequences of elements of N modulo the following equivalence relation: $\{x_n\} \sim \{y_n\}$ if, for each

$m \in \mathbb{N}$, there exists an n_0 such that $x_n - y_n \in M_m$ for all $n \geq n_0$).

Proof. To prove (i), let $P = \Pi M_i$ and let $M \subset P$ be the submodule consisting of elements $(x_i) \in P$ such that $f_i^{i+1}(x_{i+1}) = x_i$. Let $f_i : M \longrightarrow M_i$ be the projection $p_i : P \longrightarrow M_i$ restricted to M. Clearly, $f_i^{i+1} \circ f_{i+1} = f_i$. Now, let M' be given together with maps g_i. By definition of P, there exists a unique map $g : M' \longrightarrow P$ such that $g_i = p_i \circ g$. Since $f_i^{i+1} \circ g_{i+1} = g_i$, it follows that $g(M') \subset M$. Hence, M is the projective limit of $\{M_i, f_i^{i+1}\}$.

To prove (ii), let $\tilde{N} = \varprojlim N/N_n$, $x' \in \tilde{N}$, $x' = (x_n')$. For each n, choose $x_n \in N$ representing x_n'. If $m \geq n$, then $x_n \equiv x_m \bmod N_n$, so (x_n) is a Cauchy sequence in N. If $y_n \in N_n$ also represents x_n', then $y_n - x_n \in N_n$ for each n; so, $x' \longmapsto (x_n)$ is a well-defined map $\tilde{N} \longrightarrow \hat{N}$. If $(x_n) = 0$, then $(x_n) \longrightarrow 0$ in N; it follows that $x_n \in N_n$ for all n and that $x' = 0$. Finally, given a Cauchy sequence (y_n), inductively choose a subsequence (x_n) such that $x_{n+1} - x_n \in N_n$ for each n. Let $x_m' \in N_m$ be the residue class of x_m. Then $(x_m') \longmapsto (y_m)$.

Remark (1.8). - If an A-module M has two filtrations (M_n) and (M_n') such that for each n there exists an m such that $M_n \subset M_m'$ and for each m' there exists an n' such that $M_{m'}' \subset M_{n'}$, then both filtrations induce the same topology on M; hence, by (1.7), the separated completions are equal.

In particular, let q and q' be ideals of A such that $q' \subset q$ and $q^n \subset q'$ for some n. Then the q-adic and the q'-adic topologies on A and M are the same, so the corresponding separated completions coincide.

<u>Lemma (1.9)</u>. - Let

$$\begin{array}{ccccccccc}
0 & \longrightarrow & A_n & \xrightarrow{f_n} & B_n & \xrightarrow{g_n} & C_n & \longrightarrow & 0 \\
 & & \uparrow u_n^{n+1} & & \uparrow v_n^{n+1} & & \uparrow w_n^{n+1} & & \\
0 & \longrightarrow & A_{n+1} & \longrightarrow & B_{n+1} & \longrightarrow & C_{n+1} & \longrightarrow & 0
\end{array}$$

be a projective system of exact sequences of abelian groups. Then:

(i) The sequence

$$0 \longrightarrow \varprojlim A_n \xrightarrow{f} \varprojlim B_n \xrightarrow{g} \varprojlim C_n$$

is exact.

(ii) If u_n^{n+1} is surjective for each $n \geq 1$, then g is surjective.

<u>Proof</u>. The first assertion follows immediately from (1.7,(i)). Given $c \in \varprojlim C_n$, take $b_n' \in B_n$ such that $g_n(b_n') = c_n$. Construct $b \in \varprojlim B_n$ such that $g(b) = c$ inductively as follows: Let $b_0 = b_0'$; given b_n such that $v_{n-1}^{n}(b_n) = b_{n-1}$, and $g_n(b_n) = c_n$, note that $g_n(v_n^{n+1}(b_{n+1}')-b_n) = 0$. Hence, there exists $a_n \in A_n$ such that $f_n(a_n) = v_n^{n+1}(b_{n+1}')-b_n$. By hypothesis, there exists $a_{n+1} \in A_{n+1}$ such that $u_n^{n+1}(a_{n+1}) = a_n$. Let $b_{n+1} = b_{n+1}' - f_{n+1}(a_{n+1})$. Then $b \in \varprojlim B_n$ and $g(b) = c$.

<u>Proposition (1.10)</u>. - Let A be a filtered ring and M a filtered A-module. Then $M/M_n = \hat{M}/\hat{M}_n$ and, hence, $gr(M) = gr(\hat{M})$.

<u>Proof</u>. For a fixed integer n, the filtration (M_m) induces filtrations $(M_n \cap M_m)$ of M_n and $(M_n + M_m/M_n)$ on M/M_n. By (1.9),

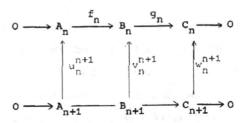

the sequence $0 \longrightarrow \hat{M}_n \longrightarrow \hat{M} \longrightarrow (M/M_n)^\wedge \longrightarrow 0$ is exact. However, since M/M_n is discrete, it follows that M/M_n is itself complete.

Definition (1.11). - Let A be a noetherian ring, q an ideal of A and M a finite A-module. A filtration (M_n) is said to be q-good if there exists a positive integer n_0 such that for each $n \geqslant n_0$, $M_{n+k} = q^k M_n$ for all $k \geqslant 0$.

Proposition (1.12). - Let A be a noetherian ring, q an ideal of A and M a filtered A-module of finite type. The following conditions on the filtration (M_n) are equivalent:

(i) The filtration (M_n) is q-good

(ii) There exists an integer n_0 such that $M_{n+1} = qM_n$ for all $n \geqslant n_0$.

(iii) $gr(M)$ is a $gr_q^*(A)$-module of finite type.

Proof. The equivalence of (i) and (ii) is trivial. If (i) holds, then $gr(M)$ is generated by $\bigoplus_{m=0}^{n_0} M_n$ over $gr(A)$; since M is of finite type over A, it follows that $gr(M)$ is of finite type over $gr_q^*(A)$. If (iii) holds, let $x_1, \ldots, x_m$ be homogeneous generators of $gr(M)$. Then, clearly, for $n \geqslant \sup\{\deg(x_i)\}$, we have $M_{n+1} = qM_n$.

Remark (1.13). - Let A be a ring and q an ideal of A. Suppose A/q is noetherian and q is finitely generated. Then $gr_q^*(A)$ is a finitely generated (A/q)-algebra; hence, $gr_q^*(A)$ is noetherian.

Theorem (1.14) (Artin-Rees). - Let A be a noetherian ring, q an ideal of A, M an A-module of finite type and N a submodule of M. Then the filtration induced on N by the q-adic filtration of M is q-good; i.e., there exists an integer n_0 such that for

$k \geq 0$

$$N \cap q^{n+k}M = q^k(N \cap q^n M) \text{ for all } k \geq 0.$$

Proof. The map $N\cap q^nM/N\cap q^{n+1}M \longrightarrow q^nM/q^{n+1}M$ is injective; hence, $gr(N) \longrightarrow gr(M)$ is injective. Since $gr(M)$ is of finite type by (1.12) and $gr(A)$ is noetherian by (1.13), $gr(N)$ is of finite type and the assertion follows from (1.12).

Theorem (1.15) (Krull intersection theorem). Let A be a noetherian ring, q an ideal of A and M a finite A-module. Then $x \in \cap q^nM$ if and only if there exists $d \in q$ such that $dx = x$. In particular, $\cap q^nM = 0$ (or equivalently, $M \longrightarrow \hat{M}$ is injective) if and only if, whenever $dx = x$ where $d \in q$ and $x \in M$, then $x = 0$.

Proof. Let $N = \cap q^nM$. By (1.14), there exists an integer k such that $q^nM\cap N = q^{n-k}(q^kM\cap N)$ for $n > k$; hence, $qN = N$. Now, the assertion follows from the next lemma.

Lemma (1.16). - Let A be a ring, N a finite A-module and q an ideal of A. Then $N = qN$ if and only if there exists $d \in q$ such that $(1-d)N = 0$.

Proof. Let $x_1,\dots,x_s$ generate N. If $N = qN$, then there exist $a_{ij} \in q$ such that $x_i = \Sigma a_{ij}x_j$. If $1-d = \det\|\delta_{ij}-a_{ij}\|$, then $d \in q$ and $(1-d)x_i = 0$, $1 \leq i \leq s$. The converse is trivial.

Proposition (1.17). - Let A be a noetherian ring, q an ideal of A and M a finite A-module. Then the additive functor $M \longmapsto \hat{M} = \varprojlim M/q^nM$ is exact.

Proof. exact sequence of of A-modules

$$0 \longrightarrow M' \longrightarrow M \longrightarrow M'' \longrightarrow 0$$

induces an exact sequence

$$0 \longrightarrow M'/(M' \cap q^n M) \longrightarrow M/q^n M \longrightarrow M''/q^n M'' \longrightarrow 0$$

for each positive integer n. By the Artin-Rees lemma (1.14) and by (1.8), the separated completion of $\{M'/M' \cap q^n M\}$ is $(M')\hat{}$. The conclusion now follows from (1.9).

<u>Theorem (1.18)</u>. - Let A be a noetherian ring, q an ideal of A and M a finite A-module. Then the canonical map $M \otimes_A \hat{A} \longrightarrow \hat{M}$ is an isomorphism.

<u>Proof</u>. By (1.17), an exact sequence

$$A^i \longrightarrow A^j \longrightarrow M \longrightarrow 0$$

yields a commutative diagram with exact rows.

$$\begin{array}{ccccccc}
A^i \otimes_A \hat{A} & \longrightarrow & A^j \otimes_A \hat{A} & \longrightarrow & M \otimes_A \hat{A} & \longrightarrow & 0 \\
\downarrow f & & \downarrow g & & \downarrow h & & \\
\hat{A}^i & \longrightarrow & \hat{A}^j & \longrightarrow & \hat{M} & \longrightarrow & 0.
\end{array}$$

Since f and g are clearly isomorphisms, the five lemma implies that h is an isomorphism.

<u>Proposition (1.19)</u>. - Let A be a noetherian ring, q and I ideals of A and M a finite A-module. Filter A and M q-adically. Then, $I\hat{M} = (IM)\hat{} = \hat{I}\hat{M}$ and, hence, $\hat{M}/I\hat{M} = (M/IM)\hat{}$. In particular, $M/q^n M = \hat{M}/q^n\hat{M} = \hat{M}/\hat{q}^n\hat{M}$, and $gr_q(M) = gr_q(\hat{M}) = gr_{\hat{q}}(\hat{M})$.

<u>Proof</u>. Consider the commutative diagram

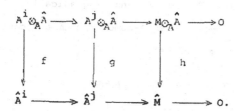

$$\begin{array}{ccc}
I\otimes_A M\otimes_A \hat{A} & \xrightarrow{\quad u \quad} & I\otimes_A \hat{M} \\
\downarrow & \overset{v}{\searrow} & \downarrow \\
IM\otimes_A \hat{A} & \xrightarrow{\quad w \quad} & \hat{M}
\end{array}$$

By (1.18), u is an isomorphism, so the image of v is $I\hat{M}$. On the other hand, by (1.17) and (1.18), w is an injection with image $(IM)^{\wedge}$, so the image of v is $(IM)^{\wedge}$. Consequently, $I\hat{A} = \hat{I}$ and $\hat{I}\hat{M} = I\hat{A}\hat{M} = I\hat{M}$; whence, by (1.17), the first assertion. The second assertion now follows from (1.10).

<u>Lemma (1.20)</u>. - Let A be a noetherian ring and B a noetherian A-algebra. Let q be an ideal of A and q' an ideal of B such that $qB \subset q' \subset \mathrm{rad}(B)$. Let M be a finite A-module and N a finite B-module. Filter A and M q-adically; B and N q'-adically. Let $\varphi : M \longrightarrow N$ be an A-homomorphism and consider the commutative diagram that φ induces:

$$\begin{array}{ccc}
M/qM & \xrightarrow{\quad \varphi'' \quad} & N/qN \\
\downarrow{\scriptstyle \alpha} & & \downarrow{\scriptstyle \beta} \\
\hat{M}/q\hat{M} & \xrightarrow{\quad \varphi' \quad} & \hat{N}/q\hat{N}.
\end{array}$$

Then:

(i) If $\hat{\varphi}$ is surjective, then β is bijective and φ'' is surjective.

(ii) If β and φ'' are surjective, then $\hat{\varphi}$ is surjective.

<u>Proof</u>. If $\hat{\varphi}$ is surjective, then φ' is surjective; so β is surjective. Since $q' \subset \mathrm{rad}(B)$, it follows from (1.15) and (1.19) that β is injective, whence, (i).

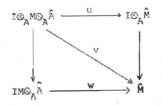

If β and φ'' are surjective, then φ' is surjective. Hence, $\hat{N} = \hat{\varphi}(\hat{M}) + q\hat{N}$. So, $q^n\hat{N} = \hat{\varphi}(q^n\hat{M}) + q^{n+1}\hat{N}$ for all $n \geq 0$, and we are reduced to proving the following lemma.

Lemma (1.21). - Let A be a ring and $u : M \longrightarrow N$ a homomorphism of filtered A-modules. Suppose M is complete, N is separated and $gr^*(u)$ is surjective. Then u is surjective and N is complete.

Proof. Let r be an integer and let $y \in N_r$. We shall construct a sequence (x_k) of elements of M_r such that $x_{k+1} \equiv x_k \bmod M_{r+k}$ and $u(x_k) \equiv y \bmod N_{r+k}$. Let $x_0 = 0$. Suppose x_k has been constructed. Then $u(x_k) \equiv y \bmod N_{r+k}$; so, by hypothesis, there exists $t_k \in M_{r+k}$ such that $u(t_k) \equiv u(x_k) - y \bmod N_{r+k+1}$. Let $x_{k+1} = x_k - t_k$ and x be a limit of the Cauchy sequence (x_k). Since M_r is closed, $x \in M_r$, and, since N is separated, $u(x) = \lim u(x_k)$ is equal to y. Therefore, $u(M_r) = N_r$; hence, u is surjective and the topology on N is the quotient of the topology on M.

Proposition (1.22). - Let A be a ring and q an ideal of A. Suppose A/q is noetherian and q is finitely generated. Then $\hat{A} = \varprojlim A/q^r$ is noetherian.

Proof. Let I be an ideal of $\hat{A}$. By (1.19), $gr_q^*(\hat{A}) = gr_q^*(A)$; hence by (1.13), $gr_q^*(I)$ is finitely generated. Let $x_1,\ldots,x_s$ be elements of I whose images $x_i' \in gr_q^{r_i}(A)$ generate $gr_q^*(I)$. Filter $E = \hat{A}^s$ by $E_r = \bigoplus_{i=1}^{s} \hat{A}_{r-r_i}$. Then $gr^*(E) = gr^*(A)^s$. Define $u : E \longrightarrow I$ by $u((a_i)) = \Sigma a_i x_i$. Then $gr(u)$ is surjective; so, by (1.21), I is finitely generated.

Lemma (1.23). - Let A be a ring. q an ideal of A, $\hat{A} = \varprojlim A/q^n$

and $\hat{q} = \varprojlim q/q^n$. Then $q\hat{A} \subset \hat{q} \subset \mathrm{rad}(\hat{A})$.

Proof. Suppose $x \in \hat{q}$. Then $x^n \in \hat{q}^n$, so Σx^n converges. Hence, for all $x \in \hat{q}$, $1/(1-x) = \Sigma x^n \in \hat{A}$. Therefore $\hat{q} \subset \mathrm{rad}(\hat{A})$.

Proposition (1.24). Let A be a ring and q an ideal of A . The map $m \mapsto \hat{m}$ induces a bijection from the set of maximal ideals of A containing q to the set of all maximal ideals of $\hat{A}$. Hence, if A is local (resp. semi-local), then $\hat{A}$ is local (resp. semi-local).

Proof. By (1.10), $A/q = \hat{A}/\hat{q}$. Hence, the assertion results from (1.23).

2. Support of a sheaf

Definition (2.1). - Let X be a ringed space and F an O_X-Module. The set of points $x \in X$ such that $F_x \neq 0$ is called the support of F and is denoted Supp(F). If A is a ring and M is an A-module, the support of M , denoted Supp(M) , is defined as $\mathrm{Supp}(\tilde{M}) \subset X = \mathrm{Spec}(A)$.

Remark (2.2). - Let X be a ringed space and $0 \to F' \to F \to F'' \to 0$ an exact sequence of O_X-Modules. Then, clearly, $\mathrm{Supp}(F) = \mathrm{Supp}(F') \cup \mathrm{Supp}(F'')$.

Proposition (2.3). - Let X be a local ringed space and F,F' O_X-Modules of finite type. Then Supp(F) is closed in X and

$$\mathrm{Supp}(F \otimes F') = \mathrm{Supp}(F') \cap \mathrm{Supp}(F) .$$

Proof. Since the support of a section is closed and F is of finite type, Supp(F) is closed. The second assertion results from the following lemma.

Lemma (2.4). - Let A be a local ring and M,N two nonzero A-modules of finite type. Then $M\otimes_A N$ is nonzero.

Proof. Let m be the maximal ideal of A. Then, by Nakayama's lemma, M/mM and N/mN are nonzero vector spaces over the field A/m; hence, their tensor product

$$(M/mM) \otimes_{A/m} (N/mN) = (M \otimes_A N) \otimes_A A/m$$

is nonzero.

Proposition (2.5). - Let X be a scheme, F a quasi-coherent O_X-Module of finite type and J the annihilator of F. Then $\mathrm{Supp}(F)$ is the underlying point-set of the subscheme $V(J)$ defined by J.

Proof. We may assume X is affine with ring A and $F = \tilde{M}$ where M is an A-module of finite type. Let $x_1,\dots,x_m$ be generators of M and I_i the annihilator of x_i. Then $V(J) = \cup V(I_i)$. On the other hand, $\mathrm{Supp}(M) = \cup \mathrm{Supp}(Ax_i) = \cup \mathrm{Supp}(A/I_i)$ and it is clear that $\mathrm{Supp}(A/I_i) = V(I_i)$, whence the assertion.

Corollary (2.6). - Let X be a scheme, J a sheaf of ideals and F a quasi-coherent O_X-Module of finite type. Then $\mathrm{Supp}(F/JF) =$ $= \mathrm{Supp}(F) \cap V(J)$.

Lemma (2.7). - Let $f : X \longrightarrow Y$ be a morphism of schemes and F an O_Y-Module of finite type. Then $\mathrm{Supp}(f^*F) = f^{-1}(\mathrm{Supp}(F))$.

Proof. If $x \in \mathrm{Supp}(f^*F)$, then $F_{f(x)} \otimes_{O_{f(x)}} O_x \neq 0$ and $x \in f^{-1}(\mathrm{Supp}(F))$. Since $O_{f(x)} \longrightarrow O_x$ is a local homomorphism, $O_x/m_{f(x)} O_x \neq 0$, so, if $F_{f(x)} \neq 0$, then, by (2.4), $F_{f(x)} \otimes_{O_{f(x)}} O_x \neq 0$ and $x \in \mathrm{Supp}(f^*F)$.

Proposition (2.8) (Weak Nullstellensatz). - Let A be a ring, M a finite A-module and $f \in A$. Then the homothety $f : M \longrightarrow M$ is nilpotent if and only if f lies in every prime of $\mathrm{Supp}(M)$. In particular, the nilradical of A (i.e., the set of all nilpotent elements of A) is the intersection of all (minimal) primes of A.

Proof. The homothety $f : M \longrightarrow M$ is nilpotent if and only if $M_f = 0$; hence, if and only if $\emptyset = \mathrm{Supp}(M_f) = \mathrm{Supp}(M) \cap D(f)$ where $D(f)$ is the set of primes not containing f.

3. Primary decomposition

Definition (3.1). - Let A be a ring and M an A-module. A prime ideal p of A is said to be associated to M if there exists an element $x \in M$ such that p is the annihilator of x. Let $\mathrm{Ass}(M)$ or $\mathrm{Ass}_A(M)$ denote the set of associated primes of M and let $\mathrm{Ann}(x)$ denote the annihilator of x. If I is an ideal of A, the primes of $\mathrm{Ass}(A/I)$ are called the essential primes of I. If X is a scheme and F is an O_X-Module, then $\mathrm{Ass}(F)$ is defined as the set of points $x \in X$ such that $m_x \in \mathrm{Ass}(F_x)$.

Remark (3.2). - Let A be a ring and M an A-module. It is clear that a prime p of A is associated to M if and only if there exists an injection $A/p \longrightarrow M$. In particular, if N is a submodule of M, then $\mathrm{Ass}(N) \subset \mathrm{Ass}(M)$. Furthermore, $\mathrm{Ass}(A/p)$ contains only the prime p and $p = \mathrm{Ann}(x)$ for all nonzero $x \in A/p$.

Proposition (3.3). - Let A be a noetherian ring and M an A-module. Then $M = 0$ if (and only if) $\mathrm{Ass}(M) = \emptyset$.

Proof. If $M \neq 0$, let I be an ideal of A which is maximal among ideals of the form $\mathrm{Ann}(x)$ for nonzero elements x of M. Since $x \neq 0$, $I \neq A$. Suppose $b, c \in A$, $bc \in I$. If $cx \neq 0$, then

$b \in \mathrm{Ann}(cx)$ and $I \subset \mathrm{Ann}(cx)$. By maximality, we have $I = \mathrm{Ann}(cx)$ and hence $b \in I$. Therefore, I is prime and $I \in \mathrm{Ass}(M)$.

Corollary (3.4). - Let A be a noetherian ring, M an A-module and $a \in A$. Then the homothety $M \xrightarrow{a} M$ is injective if and only if a does not belong to any associated prime of M.

Proof. If a belongs to an associated prime, then clearly the homothety is not injective. Conversely, suppose $ax = 0$ for some nonzero $x \in M$. Since $Ax \neq 0$, there exists $p \in \mathrm{Ass}(Ax)$ by (3.3). Then $p \in \mathrm{Ass}(M)$ and $p = \mathrm{Ann}(bx)$ for some $b \in A$. Since $abx = 0$, it follows that $a \in p$.

Corollary (3.5). - The set of zero divisors of a noetherian ring A is the union of the associated primes of A.

Lemma (3.6). - Let A be a ring, M an A-module and N a submodule of M. Then

$$\mathrm{Ass}(M) \subset \mathrm{Ass}(N) \cup \mathrm{Ass}(M/N).$$

Proof. Let $p \in \mathrm{Ass}(M)$, E the image of the corresponding map $A/p \longrightarrow M$ and $F = E \cap N$. If $F = 0$, then E is isomorphic to a submodule of M/N; hence, $p \in \mathrm{Ass}(M/N)$. If $F \neq 0$ and x is a nonzero element of F, then $\mathrm{Ann}(x) = p$ by (3.2). Hence $p \in \mathrm{Ass}(F) \subset \mathrm{Ass}(N)$.

Theorem (3.7). - Let A be a noetherian ring and M a finite A-module. Then:

(i) There exists a filtration $M = M_0 \supset \dots \supset M_n = 0$ such that $M_i/M_{i+1} \cong A/p_i$ where p_i is a prime of A.

(ii) For any such filtration $\mathrm{Ass}(M) \subset \{p_0, \dots, p_{n-1}\} \subset \mathrm{Supp}(M)$. In particular, $\mathrm{Ass}(M)$ is finite.

Proof. To prove (i), let N be a maximal submodule of M having such a filtration. If $M/N \neq 0$, then, by (3.3) M/N contains a submodule N'/N isomorphic to A/p for some prime p of A, contradicting maximality. Hence $M = N$.

The first inclusion of the second assertion follows immediately from (3.2) and (3.6). Since $p_i \in \mathrm{Supp}(A/p_i)$, the second assertion follows from (2.2).

Lemma (3.8). - Let A be a ring and M an A-module. If Ψ is a subset of $\mathrm{Ass}(M)$, then there exists a submodule N of M such that $\mathrm{Ass}(N) = \mathrm{Ass}(M) - \Psi$ and $\mathrm{Ass}(M/N) = \Psi$.

Proof. By Zorn's lemma, there exists a maximal submodule N of M such that $\mathrm{Ass}(N) \subset \mathrm{Ass}(M) - \Psi$. By (3.6), it suffices to show that $\mathrm{Ass}(M/N) \subset \Psi$. Let $p \in \mathrm{Ass}(M/N)$; then M/N contains a submodule N'/N isomorphic to A/p. By (3.2) and (3.6), $\mathrm{Ass}(N') \subset \mathrm{Ass}(N) \cup \{p\}$. Since N is maximal, $p \in \Psi$.

Proposition (3.9). - Let A be a noetherian ring, S a multiplicative set, Φ the set of primes not intersecting S and M an A-module. Then the map $p \longmapsto S^{-1}p$ is a bijection from $\mathrm{Ass}_A(M) \cap \Phi$ to $\mathrm{Ass}_{S^{-1}A}(S^{-1}M)$.

Proof. The map $p \longmapsto S^{-1}p$ is a bijection from Φ to the set of primes of $S^{-1}A$. Furthermore, if $A/p \longrightarrow M$ is injective, then $S^{-1}(A/p) = S^{-1}A/S^{-1}p \longrightarrow S^{-1}M$ is injective; so, if $p \in \mathrm{Ass}(M) \cap \Phi$, then $S^{-1}p \in \mathrm{Ass}(S^{-1}M)$.

Let $S^{-1}p \in \mathrm{Ass}(S^{-1}M)$; there exist $x \in M$ and $t \in S$ such that $S^{-1}p = \mathrm{Ann}(x/t)$. Since p is finitely generated, there exists an element $s \in S$ such that $p \subset \mathrm{Ann}(sx)$. Moreover, if $bsx = 0$, then $b/1 \in S^{-1}p$ and, hence, $b \in p$. Thus, $p = \mathrm{Ann}(sx)$ and the proof is complete.

Corollary (3.10). - Let A be a noetherian ring and M an A-module. Then $\mathrm{Supp}(M) = \cup V(p)$ as p runs through $\mathrm{Ass}(M)$.

Proof. By (3.3), $M_p \neq 0$ if and only if $\mathrm{Ass}_{A_p}(M_p) \neq \emptyset$. However, by (3.9), $\mathrm{Ass}_{A_p}(M_p) \neq \emptyset$ if and only if there exists $q \in \mathrm{Ass}(M)$ such that $q \cap (A-p) = \emptyset$; i.e., if and only if $p \supset q$ for some $q \in \mathrm{Ass}(M)$.

Remark (3.11). - Let A be a noetherian ring and M an A-module. The minimal primes of $\mathrm{Ass}(M)$ are called the minimal (or isolated) primes of M and, by (3.10), they correspond to the maximal points of $\mathrm{Supp}(M)$. Those primes of $\mathrm{Ass}(M)$ which are not minimal are called embedded primes.

Let X be a locally noetherian scheme and F an O_X-Module. A prime cycle of F is defined as a closure in X of a point $x \in \mathrm{Ass}(F)$. An embedded prime cycle of F is defined as a prime cycle which is properly contained in another prime cycle of F. The embedded prime cycles of O_X are often called the embedded components of X.

Definition (3.12). - Let A be a noetherian ring, M an A-module and Q a submodule of M. If $\mathrm{Ass}(M/Q)$ consists of a single element p, then Q is said to be p-primary with respect to M.

Definition (3.13). - Let A be a noetherian ring, M an A-module and N a submodule of M. A primary decomposition of N in M is defined as a finite family $\{Q_i\}$ of submodules of M which are primary with respect to M and such that $N = \cap Q_i$. A primary decomposition is said to be irredundant if it satisfies the following two conditions:

(a) $\bigcap_{j \neq i} Q_j \not\subset Q_i$ for any i.

(b) If p_i is the prime corresponding to Q_i, then $p_i \neq p_j$ whenever $i \neq j$.

Theorem (3.14). - Let A be a noetherian ring, M a finite A-module and N a submodule of M. Then there exists a primary decomposition of N in M, $\{Q(p)\}$, where p runs through $\mathrm{Ass}(M/N)$ and $Q(p)$ is p-primary.

Proof. Replacing M by M/N, we may assume $N = 0$. By (3.8), there exists, for each $p \in \mathrm{Ass}(M)$, a submodule $Q(p)$ of M such that $\mathrm{Ass}(M/Q(p)) = \{p\}$ and $\mathrm{Ass}(Q(p)) = \mathrm{Ass}(M) - \{p\}$. Let $P = \cap Q(p)$. Then $\mathrm{Ass}(P) \subset \mathrm{Ass}(Q(p))$ for all $p \in \mathrm{Ass}(M)$; hence, $\mathrm{Ass}(P) = \emptyset$. Thus, by (3.3), $P = 0$.

Proposition (3.15). - Let A be a noetherian ring, M an A-module and N a submodule of M. Let $\{Q_i\}$ be a primary decomposition of N in M and p_i the prime corresponding to Q_i. Then $\mathrm{Ass}(M/N) \subset \{p_i\}$ and the decomposition is irredundant if and only if $\mathrm{Ass}(M/N) = \{p_i\}$ and the p_i are distinct. Consequently, if M is of finite type, then the associated primes of M/N are precisely the associated primes of the M/Q_i appearing in an irredundant decomposition of N in M.

Proof. Since $N = \cap Q_i$, there is an injection $M/N \longrightarrow \oplus M/Q_i$. So, by (3.2) and (3.6), $\mathrm{Ass}(M/N) \subset \{p_i\}$ and, if equality holds and the p_i are distinct, $\bigcap_{j \neq i} Q_j \not\subset Q_i$ for any i.

If $\{Q_i\}$ is irredundant, let $P_i = \bigcap_{j \neq i} Q_j$. Then $P_i \cap Q_i = N$, $P_i/N \cong (P_i + Q_i)/Q_i \subset M/Q_i$ and $P_i/N \subset M/N$. It follows that $p_i \in \mathrm{Ass}(P_i/N) \subset \mathrm{Ass}(M/N)$.

Remark (3.16). - Let A be a ring, S a multiplicative set, M an A-module and N a submodule of M. Then the inverse image N' of $S^{-1}N$ under the map $M \longrightarrow S^{-1}M$ is called the saturation of N with respect to S. Clearly, N' is the set of all $x \in M$ such that $sx \in N$ for some $s \in S$.

If N is p-primary and $S \cap p = \emptyset$, then the homothety $s : M/N \longrightarrow M/N$ is injective by (3.4). Therefore, the saturation of N is equal to N.

Proposition (3.17). - Let A be a noetherian ring, M an A-module, N a submodule of M and $I = \mathrm{Ass}(M/N)$. Let S be a multiplicative set, J the subset of I consisting of those primes p_j such that $S \cap p_j = \emptyset$, and N' the saturation of N with respect to S. If $\{Q_i\}$ is an irredundant primary decomposition of N, then $\{S^{-1}Q_i\}_{i \in J}$ is an irredundant primary decomposition of $S^{-1}N$ and $\{Q_i\}_{i \in J}$ is an irredundant primary decomposition of N'.

Proof. It follows easily from (3.9) and (3.15) that $\{S^{-1}Q_i\}_{i \in J}$ is an irredundant primary decomposition of $S^{-1}N$; hence, by (3.16), we conclude that $\{Q_i\}_{i \in J}$ is an irredundant primary decomposition of N'.

Corollary (3.18). - Let A be a noetherian ring, M an A-module and N a submodule of M. If p_0 is a minimal prime of M/N and $\{Q(p)\}$ is an irredundant primary decomposition of N in M, then $Q(p_0)$ is uniquely determined by N.

Proof. If $S = A - p_0$, then $Q(p_0)$ is the saturation of N with respect to S by (3.17).

4. Length and characteristic functions

Definition (4.1). - Let A be a ring and M an A-module. A filtration

$$M = M_0 \supset \ldots \supset M_n = (0)$$

is said to be a composition series if each quotient M_i/M_{i+1} is a simple A-module. By the Jordan-Hölder theorem, any two composition series of M have the same number of terms; that number, n, is called the length of M and denoted $\ell_A(M)$ or $\ell(M)$.

Remark (4.2). - Let $0 \longrightarrow M' \longrightarrow M \longrightarrow M'' \longrightarrow 0$ be an exact sequence of A-modules. Then it is easily seen that M has finite length if and only if M' and M" have finite length. In this case, we have

$$\ell(M) = \ell(M') + \ell(M'').$$

Proposition (4.3). - Let A be a noetherian ring and M a finite A-module. Then M has finite length if and only if Ass(M) (resp. Supp(M)) consists entirely of maximal ideals.

Proof. Since all simple A-modules are isomorphic to A/m for some maximal ideal m of A, the assertion follows from (3.7) and (3.10).

Definition (4.4). - Let A be a ring. An A-module M is said to be artinian if every nonempty set of submodules of M has a minimal element, (or equivalently, if every descending chain of submodules stops).

Proposition (4.5). - Let A be a ring. An A-module M has finite length if and only if it is both artinian and noetherian.

Proof. If M has finite length, then, by the Jordan-Hölder theorem, every chain of submodules has finite length; hence, M is both artinian and noetherian. Conversely, construct a filtration (M_i) of M as follows: Let $M_0 = M$ and let M_{i+1} be a maximal proper submodule of M_i. Since this descending chain stops, it is a composition series of M.

Lemma (4.6). - Let A be a ring in which 0 is a product of maximal ideals $m_1,\ldots,m_n$. Then any prime p is one of the m_i and A is both artinian and noetherian. Moreover, if the A/m_i are algebras of finite type over a field k, then A has finite k-dimension.

Proof. Since $p \supset 0 = m_1 \ldots m_n$, it follows that $p = m_i$ for some i. Let $I_j = m_1 \ldots m_j$ for $1 \leq j \leq n$. Then A has a finite filtration $I_0 \supset \ldots \supset I_n = 0$ whose quotients I_{j-1}/I_j are finite vector spaces over A/m_j. Hence, by (4.5), A is both artinian and noetherian. Moreover, if A/m_i is of finite type over k, then it has finite k-dimension by the Hilbert Nullstellensatz (III,2.7); whence, the assertion.

Theorem (4.7). - A ring A is artinian if and only if the following two conditions hold:

(i) A is noetherian.

(ii) Every prime ideal of A is maximal.

Moreover, if A is artinian, then A has only a finite number of primes and $\mathrm{rad}(A)$ is nilpotent. If, in addition, A is of finite type over a field k, then A has finite k-dimension.

Proof. Suppose A is noetherian. Then by noetherian induction, every ideal of A contains a finite product of primes. If, in addition, every prime is maximal, then 0 may be written as a product of maximal ideals. Hence, by (4.6), A is artinian.

Conversely, suppose A is artinian. Let m be the smallest product of maximal ideals of A. Let S be the set of ideals contained in m such that $Im \neq 0$. If $I \in S$ is minimal, then $m^2I = mI \neq 0$; hence, by minimality, $mI = I$. Since $m \subset \operatorname{rad}(A)$, if $I = xA$, then $I = 0$ by Nakayama's lemma. Therefore, if $x \in I$, then $xm = 0$; so $Im = 0$. Hence, S must be empty and $m = m^2 = 0$. Thus, by (4.6), A is noetherian and every prime is maximal.

Corollary (4.8). - Let A be an artinian ring and M a finite A-module. Then M has finite length and $\operatorname{Ass}(M) = \operatorname{Supp}(M)$.

Proposition (4.9). - Let A be an artinian ring and $m_1,\ldots,m_r$ the maximal ideals of A. Then:

(i) The natural map $u : A \longrightarrow \Pi A_{m_i}$ is an isomorphism.

(ii) For n sufficiently large, the natural maps $v_i: A_{m_i} \longrightarrow A/m_i^n$ are isomorphisms.

Proof Since $X = \operatorname{Spec}(A)$ is discrete, u is simply the natural isomorphism $A \xrightarrow{\sim} \Gamma(X, O_X)$.

In general, any $s \notin m_i$ becomes a unit in the local ring A/m_i^n; hence, by the universality of A_{m_i}, v_i exists. For fixed i, consider $u_i: A \longrightarrow A_{m_i}$. Clearly, there exists $s \in (\bigcap_{j \neq i} m_j^n) - m_i$ for any n. By (4.7), if $n >> 0$, then $sa = 0$ for any $a \in m_i^n$; so, u_i induces $u_i' : A/m_i^n \longrightarrow A_{m_i}$, an inverse to v_i.

Lemma (4.10). - Any polynomial $P \in \mathbb{Q}[n]$ of degree d may be expressed in the form

$$P(n) = c_d\binom{n}{d} + c_{d-1}\binom{n}{d-1} + \dots + c_0$$

where $c_i \in \mathbb{Q}$. If $P(n)$ is an integer for all large integers n, then the c_i are all integers.

<u>Proof</u>. The assertions follow easily by induction on s from the formulas

$$n^s = s!\,\binom{n}{s} + P'(n)$$

$$\binom{n+1}{s} - \binom{n}{s} = \binom{n}{s-1}$$

where P' is a polynomial of degree $s-1$.

Let $H = \oplus H_n$ be a graded ring such that H_0 is an artinian ring and H is generated over H_0 by a finite number of elements of H_1. Let $M = \oplus M_n$ be a graded H-module of finite type. Then, by (4.8), the H_0-module M_n, being of finite type, has finite length. The function $\chi(M,n) = \ell_{H_0}(M_n)$ is called the <u>Hilbert characteristic function</u> of M. If $0 \longrightarrow M' \longrightarrow M \longrightarrow M'' \longrightarrow 0$ is an exact sequence of graded H-modules of finite type, then, by (4.2),

$$\chi(M,n) = \chi(M',n) + \chi(M'',n).$$

<u>Theorem (4.11) (Hilbert)</u>. - Let H be a graded ring satisfying:

(a) H_0 is an artinian ring.

(b) H is an H_0-algebra generated by $x_1,\dots,x_r \in H_1$.

Let M be a graded H-module of finite type. Then there exists a polynomial $Q(M,n)$ of degree $\leq r-1$ such that $\chi(M,n) = Q(M,n)$ for large integers n.

<u>Proof</u>. The proof proceeds by induction on r. If $r = 0$, then $H = H_0$ and, by (4.8), M is an H-module of finite length. Therefore, $M_n = 0$ for large n and $Q(M,n) = 0$.

Assume the assertion holds for $r-1$ and let M be a graded $H_0[x_1,\dots,x_r]$-module of finite type. The exact sequence

$$0 \longrightarrow N_n \longrightarrow M_n \xrightarrow{x_r} M_{n+1} \longrightarrow R_{n+1} \longrightarrow 0$$

yields $\Delta\chi(M,n) = \chi(M,n+1)-\chi(M,n) = \chi(R,n+1)-\chi(N,n)$. Now, N and R are graded $H_0[x_1,\dots,x_{r-1}]$-modules since x_r annihilates them. Therefore, by induction, $\Delta\chi(M,n)$ coincides for all large n with polynomial $Q(R,n+1) - Q(N,n)$ of degree $\leq r-2$. Therefore, the assertion follows from (4.10).

Lemma (4.12). - Let A be a noetherian ring, M a finite A-module, q an ideal of A and (M_n) a q-good filtration of M. If M/qM has finite length, then M/M_n has finite length for all integers $n > 0$.

Proof. By (2.6), $\mathrm{Supp}(M/q^nM) = \mathrm{Supp}(M) \cap V(q^n) = \mathrm{Supp}(M/qM)$; so, by (4.3), M/q^nM has finite length. Since $M_n \supset q^nM$ for all $n > 0$, it follows that M/M_n has finite length.

Theorem (4.13) (Samuel). - Let A be a noetherian ring, M a finite A-module and q an ideal of A such that M/qM has finite length. Let (M_n) be a q-good filtration of M.

(i) There exists a unique polynomial $P_{(M_n)}$ such that $P_{(M_n)}(m) = \ell(M/M_m)$ for large m; furthermore, $P_{(M_n)}$ depends only on $\mathrm{gr}(M)$.

(ii) If q can be generated by r elements, then $\deg(P_{(M_n)}) \leq r$.

(iii) The degree and leading coefficient of $P_{(M_n)}$ are independent of the choice of filtration.

Proof. Let I be the annihilator of M, $B = A/I$ and $p = (q+I)/I$. Filter B p-adically and let $H = \mathrm{gr}(B)$. By (2.6) and (4.7), B/p is artinian and since p is finitely generated, H satisfies (a) and (b) of (4.11). Moreover, since (M_n) is q-good, gr(M) is a finite gr(B)-module by (1.12).

Hence, by (4.11), there exists a polynomial Q(gr(M),n) which coincides with $\chi(\mathrm{gr}(M),n)$ for large n. On the other hand, $\Delta \ell(M/M_n) = \ell(M/M_{n+1}) - \ell(M/M_n) = \chi(\mathrm{gr}(M),n)$; hence, it follows from (4.10) that there exists a polynomial $P_{M_n}(n)$ which coincides with $\ell(M/M_n)$ for large n.

Since $\Delta P_{M_n}(n)$ has degree $\leq r-1$, $P_{M_n}(n)$ has degree $\leq r$ by (4.10).

To prove (iii), let n_0 be an integer such that $M_{n+1} = qM_n$ for $n \geq n_0$. Then for n large, we have

$$q^{n+n_0}M \subset M_{n+n_0} = q^n M_{n_0} \subset q^n M \subset M_n .$$

Hence, for large n,

$$P_{(q^m M)}(n+n_0) \geq P_{(M_m)}(n+n_0) \geq P_{(q^m M)}(n) \geq P_{(M_m)}(n),$$

and the proof is complete.

Definition (4.14) - The polynomial $P_{(q^m M)}$ is called the Hilbert-Samuel polynomial and is usually denoted $P_q(M,n)$.

Lemma (4.15). - Let A be a noetherian ring, q an ideal of A and $0 \longrightarrow M' \longrightarrow M \longrightarrow M'' \longrightarrow 0$ an exact sequence of A-modules of finite type. If M/qM has finite length, then M'/qM' and M"/qM" have finite length and the polynomial $P_q(M,n) - P_q(M'',n) - P_q(M',n)$ has degree $\leq \deg (P_q(M',n)) - 1$.

Proof. The filtration $(M'_n) = (M' \cap q^n M)$ of M' is q-good by the Artin-Rees lemma (1.14). Since, by (4.2),

$$\ell(M/q^n M) = \ell(M''/q^n M'') + \ell(M'/M'_n)$$

the conclusion follows from (4.13,(iii)).

Chapter III - Depth and Dimension

1. Dimension theory in noetherian rings

Remark (1.1). - Let X be a topological space. The dimension of X, denoted $\dim(X)$, is defined as the supremum of all integers r such that there exists a chain of closed irreducible subsets

$$X = X_0 \supsetneq X_1 \supsetneq \cdots \supsetneq X_r.$$

If A is a ring, the dimension of $X = \mathrm{Spec}(A)$ is called the (Krull) dimension of A and is denoted $\dim(A)$. Let M be an A-module and I the annihilator of M. The dimension of M, denoted $\dim(M)$, is defined as the dimension of the ring A/I; M is said to be equidimensional if $\dim(M) = \dim(A/p)$ for all minimal essential primes p of I. If p is a prime, then the height of p is defined as the dimension of A_p. If A is noetherian and M is a finite A-module, then, by (II, 2.5), $\dim(M) = \dim(\mathrm{Supp}(M))$; by (II,3.10), $\dim(\mathrm{Supp}(M))$ is equal to the supremum of the integers $\dim(A/p)$ as p ranges over $\mathrm{Ass}(M)$ (resp. $\mathrm{Supp}(M)$).

Remark (1.2). - Let A be a semilocal noetherian ring. An ideal q of A is said to be an ideal of definition of A if the following two conditions hold:

(a) $q \subset \mathrm{rad}(A)$.

(b) A/q is an artinian ring

If $q' \supset q$ is another ideal of definition, then, by (II,4.7), $q'^m \subset q$ for some integer m.

Let A be a semilocal noetherian ring, q an ideal of definition of A and M a finite A-module. The, by (II,4.8), M/qM has finite length. Furthermore, it is clear that if $q' \subset q$ is another ideal of definition, then $P_{q'}(M,n) \leq P_q(M,n)$ and $P_q(M,n) \leq P_{q'}(M,mn)$ (II,4.13). Therefore, the degree $d(M)$ of $P_q(M,n)$ is independent of q.

Let $s(M)$ be the smallest integer r such that there exist $x_1,\ldots,x_r \in \mathrm{rad}(A)$ with $M/(x_1M + \ldots + x_rM)$ of finite length.

Lemma (1.3). - Let A be a semilocal noetherian ring and M a finite A-module. Let $x \in \mathrm{rad}(A)$ and let ${}_xM$ be the kernel of the homothety $M \xrightarrow{x} M$. Then

(i) $s(M) \leq s(M/xM) + 1$.

(ii) Let $\{p_i\}$ be the primes of $\mathrm{Supp}(M)$ such that $\dim(A/p_i) = \dim(A)$. If $x \notin \cup p_i$, then $\dim(M/xM) \leq \dim(M)-1$.

(iii) If q is an ideal of definition of A, then the polynomial $P_q({}_xM) - P_q(M/xM)$ has degree $\leq d(M)-1$.

Proof. Assertions (i) and (ii) are trivial. To prove (iii), apply (II,4.15) to the exact sequences

$$0 \longrightarrow {}_xM \longrightarrow M \longrightarrow xM \longrightarrow 0$$

$$0 \longrightarrow xM \longrightarrow M \longrightarrow M/xM \longrightarrow 0.$$

Theorem (1.4). - Let A be a semilocal noetherian ring and M a finite A-module. Then

$$\dim(M) = d(M) = s(M).$$

Proof. Step I. $\dim(M) \leq d(M)$.

If $d(M) = 0$, then M has finite length and, by (II,4.3) and (1.1) $\dim(M) = 0$.

Suppose $d(M) \geq 1$ and $p_0 \in \mathrm{Ass}(M)$ is such that $\dim(A/p_0) = \dim(M)$. Then M contains a submodule N isomorphic to A/p_0 and, by

(II,4.2), $d(N) \leq d(M)$. Thus, it suffices to prove Step I for $M = A/p_0$.

Let $p_0 \subsetneqq \cdots \subsetneqq p_n$ be a chain of primes of A, If $n = 0$, then clearly $n \leq d(M)$. If $n > 0$, choose $x \in p_1 \cap \mathrm{rad}(A)$, but $x \notin p_0$. The chain $p_1 \subsetneqq \cdots \subsetneqq p_n$ belongs to Supp(M/xM); so, $n-1 \leq \dim(M/xM)$. However, ${}_xM = 0$; by (1.3), $d(M/xM) \leq d(M)-1$. Hence, Step I follows by induction on $d(M)$.

Step II. $d(M) \leq s(M)$.

Let $I = x_1A + \cdots + x_rA$ be such that $I \subset \mathrm{rad}(A)$ and M/IM has finite length. If $q = I + (\mathrm{rad}(A) \cap \mathrm{Ann}(M))$, then q is an ideal of definition of A. Indeed, $q \subset \mathrm{rad}(A)$ and $V(q) =$ $= V(I) \cap (V(\mathrm{rad}(A)) \cup \mathrm{Supp}(M))$ consists entirely of maximal ideals. Furthermore, by (II,4.13), $P_q(M,n) = P_I(M,n)$ since $I^nM = q^nM$ for all n. Again, by (II,4.13), $P_I(M,n)$ has degree $\leq r$. Therefore, $d(M) \leq s(M)$.

Step III. $s(M) \leq \dim(M)$.

The proof proceeds by induction on $n = \dim(M)$, which is finite by Step I. If $n = 0$, M has finite length by (II,4.3).

Suppose $n \geq 1$ and let $\{p_i\}$ be the primes of Supp(M) such that $\dim(A/p_i) = n$. They are not maximal since $n \geq 1$; hence, by the following lemma, there exists $x \in \mathrm{rad}(A)$ such that $x \notin p_i$ for all i. By (1.3), $s(M) \leq s(M/xM) + 1$ and $\dim(M) \geq \dim(M/xM) + 1$. By induction, $s(M/xM) \leq \dim(M/xM)$; so $s(M) \leq \dim(M)$.

Lemma (1.5). - Let A be a ring and E a subset of A which is stable under addition and multiplication; let $\{p_i\}_{i=1}^h$ be a nonempty family of ideals of A such that $p_3,\ldots,p_h$ are prime. If $E \subset \cup p_i$, then $E \subset p_i$ for some i.

Proof. The assertion is trivial for $h = 1$, so assume $h > 1$. Since $E = \cup(E \cap p_i)$, we may suppose by induction on h that there is no index j such that $E \cap p_j \subset \bigcup_{i \neq j} p_i$. For each j, choose an element $x_j \in E \cap p_j$ such that $x_j \notin p_i$ for $i \neq j$. Then $y = x_h + \prod_{j \neq h} x_j \in E$, but $y \notin p_i$ for any i.

Corollary (1.6). - Let A be a semilocal noetherian ring and M a finite A-module. Then, for each $x \in \mathrm{rad}(A)$,

$$\dim(M/xM) \geq \dim(M) - 1,$$

with equality if $x \notin p$ where p runs through the primes of $\mathrm{Supp}(M)$ such that $\dim(M) = \dim(A/p)$.

Proof. By (1.3), $s(M/xM) \geq s(M) - 1$; hence, the assertion follows from (1.4).

Corollary (1.7). - Let $\varphi : A \longrightarrow B$ be a local homomorphism of noetherian rings, m the maximal ideal of A and $k = A/m$. Then

$$\dim(B) \leq \dim(A) + \dim(B \otimes_A k).$$

Proof. Let $d = \dim(A)$ and let I be an ideal generated by d elements of m such that A/I has finite length. By (II,4.5), A/I is artinian; so, by (II,4.7), m/I is nilpotent. Hence, mB/IB is nilpotent and, thus, $\dim(B \otimes_A k) = \dim(B/IB)$. By (1.6), $\dim(B/IB) \geq \dim(B) - d$; whence, the assertion.

Corollary (1.8). - Let A be a semilocal noetherian ring and M a finite A-module. Then $\dim_A(M) = \dim_{\hat{A}}(\hat{M})$.

Proof. By (II,1.19) and (II,4.13), $d(M) = d(\hat{M})$; hence, the assertion follows from (1.4).

Corollary (1.9). - Let A be a noetherian ring, p a prime of A and n integer. The following conditions are equivalent:

(i) $ht(p) \leq n$.

(ii) There exists an ideal I of A generated by n elements such that p is a minimal (essential) prime of I.

Proof. If (ii) holds, IA_p is an ideal of definition of A_p. Hence, $ht(p) = \dim(A_p) = s(A_p) \leq n$. Conversely, if (i) holds, there exists an ideal of definition of A_p generated by n elements $\frac{x_i}{s}$ where $s \in A - p$. It follows by (II,3.9) that p is a minimal prime of $I = x_1A + \cdots + x_nA$.

Remark (1.10). - With $n = 1$, (1.9) is known as Krull's principal ideal theorem.

2. Dimension theory in algebras of finite type over a field.

Lemma (2.1). - Let A,B be domains and suppose B is integral over A. Then B is a field if and only if A is a field.

Proof. Suppose B is a field and let a be a nonzero element of A. Since $1/a \in B$, it satisfies an equation $(1/a)^n + a_{n-1}(1/a)^{n-1} + \dots + a_0 = 0$ with $a_i \in A$. Then $1/a = -(a_{n-1} + aa_{n-2} + \dots + a^{n-1}a_0)$ and, consequently, $1/a \in A$.

Conversely, suppose A is a field and let b be a nonzero element of B. Then b satisfies an equation $b^n + a_{n-1}b^{n-1} + \dots + a_0 = 0$ with $a_i \in A$ and $a_0 \neq 0$. Hence, $1/b = -((a_1/a_0) + \dots + (a_{n-1}/a_0)b^{n-2} + (1/a_0)b^{n-1}) \in B$.

Proposition (2.2) (Cohen-Seidenberg). - Let A be a subring of B and p a prime of A. Suppose B is integral over A.

(i) If P is a prime of B lying over p, then P is maximal if and only if p is maximal.

(ii) If $P' \supset P$ are primes of B lying over p, then $P = P'$.

(iii) If p is any prime of A, there exists a prime P of B lying over p.

Proof. Assertion (i) follows from (2.1) applied to A/p and B/P. To prove (ii) and (iii), replace A by $S^{-1}A$ and B by $S^{-1}B$ where $S = A - p$; then, A is local with maximal ideal p. Now, (i) implies (ii) and that, if P is any maximal ideal of B, then $p = P \cap A$, completing the proof.

Lemma (2.3). - Let A be a domain integrally closed in its quotient field K. Let L be a finite normal extension of K, B the integral closure of A in L, G the group of K-automorphisms of L and p a prime of A. Then G operates transitively on the primes of B lying over p.

Proof. Let P, P' be primes of B lying over p. If $g \in G$, the prime gP lies over p and, by (2.2), it suffices to show that $P' \subset gP$ for some $g \in G$. Let $b \in P'$ and let $a = \Pi g(b)$. Then $a^q \in K$ where q is a power of the characteristic of K. Since A is integrally closed, $a^q \in A$ and thus $a^q \in p$. Hence, there exists an automorphism g such that $g(b) \in P$, and $b \in g^{-1}P$. Hence, $P' \subset \cup gP$; so, by (1.5), $P' \subset gP$ for some g.

Proposition (2.4) (Cohen-Seidenberg). - Let B be a domain, A a subdomain of B, $p \subsetneq p'$ primes of A, and P' a prime of B lying over p'. Suppose A is integrally closed and B is a finite A-module. Then there exists a prime $P \subsetneq P'$ lying over p.

Proof. Let K be the quotient field of A, L a finite normal extension of K containing B, and C the integral closure of A in L. By (2.2), there exist a prime Q' of C lying over P' and a chain $Q \subsetneq Q''$ of primes of C lying over $p \subsetneq p'$. By (2.3), there exists a K-automorphism g of L such that $gQ'' = Q'$. If $P = gQ \cap B$, then P is the required prime.

Theorem (2.5). (Noether normalization lemma). - Let k be a field, A a k-algebra of finite type and $I_1 \subset \ldots \subset I_r$ a sequence of ideals of A with $I_r \neq A$. Then there exist elements $t_1, \ldots, t_n$ of A, algebraically independent over k, such that:

(a) A is integral over $B = k[t_1, \ldots, t_n]$.

(b) For each i, $1 \leq i \leq r$, there exists an integer $h(i) \geq 0$ such that $I_i \cap B$ is generated by $\{t_1, \ldots, t_{h(i)}\}$.

Proof. A is a quotient of a polynomial algebra $A' = k[T_1, \ldots T_m]$ and clearly we may assume $A = A'$. The proof proceeds by induction on r.

Step I. Suppose $r = 1$ and I_1 is a principal ideal generated by a nonzero element t_1. By assumption, $t_1 = P(T_1, \ldots, T_m) \notin k$ where $P = \Sigma a_{(j)} T^{(j)} \in k[T_1, \ldots, T_m]$. We are going to choose positive integers s_i such that A is integral over $B = k[t_1, \ldots, t_m]$ where $t_i = T_i - T_1^{s_i}$, $2 \leq i \leq m$. To do this, it will suffice to show that T_1 is integral over B.

Now T_1 satisfies the equation

$$t_1 - \sum a_{(j)} T_1^{j_1} (t_2 + T_1^{s_2})^{j_2} \ldots (t_m + T_1^{s_m})^{j_m} = 0.$$

Let $f(j) = j_1 + s_2 j_2 + \ldots + s_m j_m$. If $s_i = \ell^i$ where ℓ is an integer greater than $\deg(P)$, then the $f(j)$ are distinct. Suppose $f(j')$

is largest among the $f(j)$. Then the above equation may be written $a_{(j')}T_1^{f(j')} + \sum_{v<f(j')} Q_v(t)T_1^v$ and, hence, T_1 is integral over B.

Clearly, $t_1,\ldots,t_m$ are algebraically independent. Suppose $x \in I_1 \cap B$. Then $x = t_1x'$ where $x' \in A \cap k(t_1,\ldots,t_m)$. Furthermore, $A \cap k(t_1,\ldots,t_m) = B$ since B is integrally closed. Hence $I_1 \cap B = t_1B$ and the proof of Step I is complete.

Step II. Suppose $r = 1$ and I_1 is arbitrary. The proof proceeds by induction on m. The case $m = 0$ is trivial. We may assume $I_1 \neq 0$. Let t_1 be a nonzero element of I_1. Then $t_1 \notin k$ because $I_1 \neq A$. By Step I, there exist elements $u_2,\ldots,u_m$ such that $t_1, u_2,\ldots,u_m$ are algebraically independent and satisfy (a) and (b) with respect to A and (t_1). By induction, there exist algebraically independent elements $t_2,\ldots,t_m$ satisfying (a) and (b) with respect to $k[u_2,\ldots,u_m]$ and $I \cap k[u_2,\ldots,u_m]$. Then $t_1,\ldots,t_m$ are algebraically independent and satisfy (a) and (b) with respect to A and I_1.

Step III. Assume the theorem holds for $r-1$. Let $u_1,\ldots,u_m$ be algebraically independent elements of A satisfying (a) and (b) for the sequence $I_1 \subset \ldots \subset I_{r-1}$ and let $s = h(r-1)$. By Step II, there exist algebraically independent elements $t_{s+1},\ldots,t_m$ satisfying (a) and (b) for $k[u_{s+1},\ldots,u_m]$ and $I_r \cap k[u_{s+1},\ldots,u_m]$. If we set $t_i = u_i$ for $i \leq s$, then $t_1,\ldots,t_m$ are algebraically independent and satisfy (a) and (b) for $I_1 \subset \ldots \subset I_r$.

Theorem (2.6). - Let A be a domain of finite type over a field k.

(i) If $p_0 \subsetneq \ldots \subsetneq p_r$ is a saturated chain of primes of A, then r is equal to $\text{tr.deg}_k A$, (the transcendence degree of A over k).

(ii) $\mathrm{tr.deg}_k A = \dim(A)$

(iii) If p is any prime of A, then $\dim(A_p) + \dim(A/p) = \dim(A)$

<u>Proof</u>. Assertion (i) implies (ii) directly, (iii) by application to chains through p. To prove (i), by (2.5), choose algebraically independent elements $t_1,\ldots,t_n \in A$ such that A is integral over $B = k[t_1,\ldots,t_n]$ and $p_i' = p_i \cap B = (t_1,\ldots,t_{h(i)})$. Then $n = \mathrm{tr.deg}_k A$ and, by (2.2), $r \leq n$; since the chain is saturated, $h(r) = n$ by (2.2) and $h(i+1) = h(i)+1$ by (2.4) applied to A/p_i and $B/p_i' \cong k[t_{h(i)+1},\ldots,t_n]$. It follows that $r = h(r) = n$.

<u>Corollary (2.7) (Hilbert Nullstellensatz)</u>. - Let A be an algebra of finite type over a field k and m a maximal ideal of A. Then the field A/m is algebraic over k.

<u>Proposition (2.8)</u>. - Let k be a field and X an algebraic k-scheme. Then:

(i) A point $x \in X$ is closed if and only if $k(x)$ is a finite extension of k.

(ii) The closed points of X are dense.

<u>Proof</u>. Since a point x is closed if and only if x is closed in every affine open subset containing x, it follows that we may assume X is affine. Let A be the ring of X, m the ideal of x in A. Then x is closed if and only if m is maximal. However, by the Hilbert Nullstellensatz (2.7), m is maximal if and only if A/m is a finite field extension of k.

3. Depth

<u>Definition (3.1)</u>. - Let A be a ring and M an A-module. Let $(x_1,\ldots,x_r)$ be a sequence of elements of A and $M_i =$

$= M/(x_1M + \dots + x_iM)$. Then $(x_1,\dots,x_r)$ is said to be M-<u>regular</u> if the sequences

$$0 \longrightarrow M_i \xrightarrow{x_{i+1}} M_i$$

are exact for $0 \leq i \leq r-1$.

<u>Lemma (3.2)</u>. - Let A be a ring and M an A-module. Let x be an element of A, J an ideal of A and $I = J + xA$. If x is $gr_J^*(M)$-regular, then the surjection defined by $T \longmapsto x$,

$$\varphi : gr_J^*(M) \otimes_A (A/xA)[T] \longrightarrow gr_I^*(M),$$

is an isomorphism. Conversely, if M/JM is separated for the I-adic topology and φ is an isomorphism, then x is (M/JM)-regular.

<u>Proof</u>. Assume x is $gr_J^*(M)$-regular. Let $P_k = (gr_J^*(M) \otimes_A (A/xA)[T])_k$ and $Q_k = gr_I^k(M)$, and filter them by $(P_k)_i = \bigoplus_{j \leq k-i} gr_J^{k-j}(M) \otimes_A (A/xA)T^j$ and $(Q_k)_i = \varphi((P_k)_i)$. Then, by (II,1.5), to prove φ_k injective, it suffices to prove $\varphi_{k,i}: gr^i(P_k) \longrightarrow gr^i(Q_k)$ injective for each i since $(P_k)_{k+1} = 0$. However, $gr^i(P_k) = (J^iM/(xJ^iM+J^{i+1}M))T^{k-i}$ and $(Q_k)_{i+1}$ is the image of $R_k = J^kM + xJ^{k-1}M + \dots + x^{k-i-1}J^{i+1}M$ in $I^kM/I^{k+1}M$. Hence, it remains to show that, if $y \in J^iM$ and $x^{k-i}y \in R_k + I^{k+1}M$, then $y \in xJ^iM + J^{i+1}M$.

By (II,1.5), x is (M/J^hM)-regular for any $h > 0$. Set $h = i + 1$; since $x^{k-i}y \in J^{i+1}M + I^{k+1}M \subset J^{i+1}M + x^{k-i+1}M$, there exists $z \in M$ such that $y - xz \in J^{i+1}M$. Set $h = i$; since $y \in J^iM$ and $xz \in J^iM$, it follows that $z \in J^iM$. Hence, $y \in xJ^iM + J^{i+1}M$ and φ is injective.

Conversely, let $\varphi(\xi \otimes T^{k-1}) \in gr_I^{k-1}(M/JM)$ where $\xi \in M/JM$. Suppose $gr_I^*(x)(\varphi(\xi \otimes T^{k-1})) = \varphi(\xi \otimes T^k)$ is zero. Then $\xi = 0$, so by (II,1.5), x is (M/JM)-regular.

Definition (3.3). - Let A be a ring and M and A-module. A sequence $(x_1,\ldots,x_r)$ of elements of A is said to be M-quasi-regular if the canonical surjection

$$\varphi_r\colon (M/IM)[T_1,\ldots,T_r] \longrightarrow gr_I^*(M),$$

where $I = x_1A + \ldots + x_rA$, is an isomorphism.

Theorem (3.4). - Let A be a ring and M an A-module. Then an M-regular sequence $(x_1,\ldots,x_r)$ is M-quasi-regular. Conversely, if $(x_1,\ldots,x_r)$ is M-quasi-regular and if M, $M/x_1M,\ldots,M/(x_1M + \ldots + x_{r-1}M)$ are separated for the I-adic topology where $I = x_1A + \ldots + x_rA$, then $(x_1,\ldots,x_r)$ is M-regular.

Proof. Assume $(x_1,\ldots,x_r)$ is M-regular. If $r = 0$, the assertion is trivial. Proceeding by induction, assume $\varphi_{r-1}\colon (M/JM)[T_1,\ldots,T_{r-1}] \longrightarrow gr_J^*(M)$ is an isomorphism where $J = x_1A + \ldots + x_{r-1}A$. Then, since x_r is (M/JM)-regular, x_r is $gr_J^*(M)$-regular. So, by (3.2), $\varphi : gr_J^*(M) \otimes_A (A/x_rA)[T_r] \longrightarrow gr_I^*(M)$ is an isomorphism; therefore, $\varphi_r = \varphi \circ (\varphi_{r-1} \otimes \mathrm{id})$ is an isomorphism and $(x_1,\ldots,x_r)$ is M-quasi-regular.

Conversely, assume φ_r is an isomorphism. If $r = 0$, the assertion is trivial. If $r > 0$, then $\varphi_r = \varphi \circ (\varphi_{r-1} \otimes \mathrm{id})$ and φ_{r-1} is surjective; so, φ is an isomorphism. Hence, by (3.2), x_r is (M/JM)-regular. Furthermore, φ_r decomposes into surjections

$$\begin{array}{ccc}
(M/IM)[T_1,\ldots,T_r] & \xrightarrow{\quad\varphi_r\quad} & gr_I^*(M) \\
\downarrow & & \uparrow \\
gr_I^*((M/JM)[T_1,\ldots,T_{r-1}]) & \xrightarrow{\;gr_I^*(\varphi_{r-1})\;} & gr_I^*(gr_J^*(M)) .
\end{array}$$

Thus, $gr_I^*(\varphi_{r-1})$ is injective; hence, since M/JM is separated, φ_{r-1} is injective by (II,1.5). Therefore $(x_1,\ldots,x_{r-1})$ is M-quasi-regular. Since $J \subset I$, by induction $(x_1,\ldots,x_{r-1})$ is M-regular; so, the proof is complete.

Corollary (3.5). - Let A be a noetherian ring and M a finite A-module. Then elements $x_1,\ldots,x_r \in \mathrm{rad}(A)$ are M-regular if and only if they are M-quasi-regular. In particular, M-regularity does not depend on the order.

Proof. The assertion follows immediately from (II,1.15) and (3.4)

Lemma (3.6). - Let A be a ring and N a finite A-module. For each $p \in \mathrm{Supp}(N)$, there exists a nonzero A-homomorphism $\varphi : N \longrightarrow A/p$.

Proof. For $p \in \mathrm{Supp}(N)$, N_p/pN_p is a nonzero vector space over K, the quotient field of A/p. Hence, there exists a nonzero map $\varphi' : N_p/pN_p \longrightarrow K$. If $y_1,\ldots,y_n$ generate N/pN as an A/p-module, there exists $s \in A-p$ such that $s\varphi'(y_i) \in A/p$ for all i. Hence, $s\varphi'$ is nonzero and maps N/pN into A/p. Take φ to be the composition

$$N \longrightarrow N/pN \xrightarrow{s\varphi'} A/p.$$

Lemma (3.7).- Let A be a noetherian ring, I an ideal of A and M a finite A-module. Then the following conditions are equivalent:

(i) $\mathrm{Ass}(M) \cap V(I) = \emptyset$

(ii) There exists $x \in I$ which is M-regular.

(iii) $\mathrm{Hom}(N,M) = 0$ for all finite A-modules N such that $\mathrm{Supp}(N) \subset V(I)$.

(iv) $\mathrm{Hom}(N,M) = 0$ for some finite A-module N such that $\mathrm{Supp}(N) = V(I)$.

Proof. Assume (i) holds. If $p \in \mathrm{Ass}(M)$, then $I \not\subset p$. By (II,3.7), $\mathrm{Ass}(M)$ is finite; hence by (1.5), there exists $x \in I$ such that $x \notin \bigcup p$ where p runs through $\mathrm{Ass}(M)$. By (II,3.4), x is M-regular and (ii) holds.

To prove (iii)$\Longrightarrow$(iv), take $N = A/I$

We prove (iv)$\Longrightarrow$(i) by contradiction. Let $p \in \mathrm{Ass}(M) \cap V(I)$. Then, by (3.6), there exists a nonzero map $\varphi : N \longrightarrow A/p$; the composition of φ with the injection $A/p \longrightarrow M$, (II,3.2), is a nonzero map $N \longrightarrow M$.

The implication (ii)$\Longrightarrow$(iii) is the case $r = 1$ in the implication (iv)$\Longrightarrow$(i) below.

Proposition (3.8). - Let A be a noetherian ring, I an ideal of A, and M a finite A-module. For any integer r, the following conditions are equivalent:

(i) $\mathrm{Ext}_A^q(N,M) = 0$ for all $q < r$ and all finite A-modules N such that $\mathrm{Supp}(N) \subset V(I)$.

(ii) $\mathrm{Ext}_A^q(N,M) = 0$ for all $q < r$ and some finite A-module N such that $\mathrm{Supp}(N) = V(I)$.

(iii) Given $x_1,\dots,x_n \in I$ such that $(x_1,\dots,x_n)$ is M-regular, there exist $x_{n+1},\dots,x_r \in I$ such that $(x_1,\dots,x_r)$ is M-regular.

(iv) There exists an M-regular sequence $(x_1,\dots,x_r)$ with all $x_i \in I$.

Proof. To prove (i)$\Longrightarrow$(ii), take $N = A/I$

Assume (ii). For $r = 0$, (iii) is trivial. Assume $r \geq 1$ and that $x_1,\dots,x_n \in I$ are such that $(x_1,\dots,x_n)$ is M-regular. If

$n = 0$, use (iv)$\Longrightarrow$(ii) of (3.7) to construct x_1; hence, we may assume $n \geq 1$. If $M_1 = M/x_1M$, the sequence $0 \longrightarrow M \xrightarrow{x_1} M \longrightarrow M_1 \longrightarrow 0$ is exact and yields an exact sequence

$$\mathrm{Ext}^q_A(N,M) \longrightarrow \mathrm{Ext}^q_A(N,M_1) \longrightarrow \mathrm{Ext}^{q+1}_A(N,M).$$

Thus, (ii) implies that $\mathrm{Ext}^q_A(N,M_1) = 0$ for $q < r-1$. Furthermore, $(x_2,\ldots,x_n)$ is M_1-regular. Hence, by induction, there exist $x_{n+1},\ldots,x_r \in I$ such that $(x_2,\ldots,x_r)$ is M_1-regular. Then $(x_1,\ldots,x_r)$ is an M-regular sequence.

The implication (iii)$\Longrightarrow$(iv) is trivial.

Assume (iv) and let N be a finite A-module such that $\mathrm{Supp}(N) \subset V(I)$. Then (i) holds trivially for $r = 0$. Assume $r \geq 1$. Then the sequence $0 \longrightarrow M \xrightarrow{x_1} M \longrightarrow M_1 \longrightarrow 0$ is exact and yields the exact sequence

$$\mathrm{Ext}^q_A(N,M_1) \longrightarrow \mathrm{Ext}^{q+1}_A(N,M) \xrightarrow{u} \mathrm{Ext}^{q+1}_A(N,M).$$

By induction, $\mathrm{Ext}^q_A(N,M_1) = 0$ for $q < r-1$, so u is injective.

However, u is induced by multiplication by x_1 on M, but may be regarded as induced by multiplication by x_1 on N. Now, $x_1 \in I$ and $\mathrm{Supp}(N) \subset V(I)$; hence, by (II,2.8), $x_1 : N \longrightarrow N$ is nilpotent. Thus, u is a nilpotent injection. Therefore, $\mathrm{Ext}^{q+1}_A(N,M) = 0$.

<u>Definition (3.9)</u>. - Let A be a noetherian ring, I an ideal of A and M a finite A-module. The <u>depth</u> of M with respect to I, denoted $\mathrm{depth}_I(M)$, is defined as the supremum of all integers r such that there exists an M-regular sequence $(x_1,\ldots,x_r)$ of elements $x_i \in I$.

<u>Corollary (3.10)</u>. - Let A be a noetherian ring, I an ideal of A, M a finite A-module and x an M-regular element of I. Then $\mathrm{depth}_I(M/xM) = \mathrm{depth}_I(M) - 1$.

Remark (3.11). - Let A be a noetherian local ring, m the maximal ideal and M a finite A-module. In place of "$\mathrm{depth}_m(M)$", we usually write "$\mathrm{depth}_A(M)$" or simply "depth(M)". By (3.7), $\mathrm{depth}(M) = 0$ if and only if $m \in \mathrm{Ass}(M)$.

Definition (3.12). - Let P be a locally noetherian scheme, X a closed suscheme of P and F a coherent O_P-Module. Then the depth of F with respect to X, denoted $\mathrm{depth}_X(F)$ is the infimum of the integers $\mathrm{depth}_{O_x}(F_x)$ as x runs through X.

Proposition (3.13). - Let P be a locally noetherian scheme, X a closed subscheme of P and F a coherent O_P-Module. Then the following conditions are equivalent:

(i) $\underline{\mathrm{Ext}}^q_{O_P}(G,F) = 0$ for all $q < r$ and all coherent O_P-Modules G with $\mathrm{Supp}(G) \subset X$.

(ii) $\underline{\mathrm{Ext}}^q_{O_P}(G,F) = 0$ for all $q < r$ and some coherent O_P-Module G with $\mathrm{Supp}(G) = X$.

(iii) $\mathrm{Depth}_X(F) \geq r$.

(iv) $\mathrm{Depth}(F_x) \geq r$ for all $x \in X$.

Proof. It follows from (IV,3.2), that $\underline{\mathrm{Ext}}^q_{O_P}(G,F)_x = \mathrm{Ext}^q_{O_{P,x}}(G_x, F_x)$. Therefore, the equivalences follow from the definitions and (3.8).

Corollary (3.14). - Let P be a noetherian affine scheme with ring A, $X = V(I)$ a closed subscheme and F a coherent O_P-Module with $\Gamma(P,F) = M$. Then $\mathrm{depth}_X(F) = \mathrm{depth}_I(M)$.

Proof. Since, by (IV,3.2), $\underline{\mathrm{Ext}}^q_{O_P}(G,F)$ is quasi-coherent, (3.14) follows from (3.13) and (3.8).

Proposition (3.15). - Let A be a noetherian local ring and M a finite A-module. Then $\mathrm{depth}(M) \leq$ the infimum of $\dim(A/p)$ as

p runs through $\mathrm{Ass}(M)$. Furthermore, $\mathrm{depth}(M)$ is infinite if and only if $M = 0$. In particular, $\mathrm{depth}(M) \leq \dim(M)$ if $M \neq 0$.

Proof. We prove by induction on r that if $r \leq \mathrm{depth}(M)$, then $r \leq \dim(A/p)$ for any $p \in \mathrm{Ass}(M)$. If $0 < r \leq \mathrm{depth}(M)$, then there exists an M-regular element $x \in m$. Let $M' = M/xM$. Then the sequence $0 \longrightarrow M \xrightarrow{x} M \longrightarrow M' \longrightarrow 0$ is exact. By (3.10), $r-1 \leq \mathrm{depth}(M')$; so, by induction, $r-1 \leq \dim(A/p')$ for any p' in $\mathrm{Ass}(M')$. It now suffices to show that for each $p \in \mathrm{Ass}(M)$, there exists $p' \in \mathrm{Ass}(M') \cap V(p + xA)$. For, since $x \notin p$, $\dim(A/p) \geq \dim(A/p') + 1 \geq r$.

By (3.7), it suffices to show that $\mathrm{Hom}(A/p + xA, M') \neq 0$. However, $\mathrm{Hom}(A/p + xA, M') = \mathrm{Hom}(A/p, M')$, and the sequence

$$0 \longrightarrow \mathrm{Hom}(A/p, M) \xrightarrow{x} \mathrm{Hom}(A/p, M) \longrightarrow \mathrm{Hom}(A/p, M')$$

is exact; its first two terms are nonzero since $p \in \mathrm{Ass}(M)$. Since $x \in m$, Nakayama's lemma implies that $\mathrm{Hom}(A/p, M') \neq 0$

If $M = 0$, then clearly any sequence is M-regular and $\mathrm{depth}(M)$ is infinite. The converse now follows from (1.4) and (II,3.3). The last statement is clear, since $\dim(M)$ is the supremum of $\dim(A/p)$ as p runs through $\mathrm{Ass}(M)$, (1.1).

Proposition (3.16). - Let A, B be noetherian local rings, $\varphi : A \longrightarrow B$ a local homomorphism and M a B-module which is of finite type over A. Then $\mathrm{depth}_A(M) = \mathrm{depth}_B(M)$.

Proof. Let m be the maximal ideal of A and let $x_1, \ldots, x_r \in m$ form an M-regular sequence. Trivially, $\varphi(x_1), \ldots, \varphi(x_r)$ form an M-regular sequence in B. Let $N = M/(x_1M + \ldots + x_rM)$; by (3.10), $\mathrm{depth}_B(N) = \mathrm{depth}_B(M) - r$ and $\mathrm{depth}_A(N) = \mathrm{depth}_A(M) - r$. It follows that we may assume $\mathrm{depth}_A(M) = 0$.

Let $P = \mathrm{Hom}_A(A/m,M)$. Then P is a B-submodule of $\mathrm{Hom}_A(A,M) = M$ and, by (3.11), $P \neq 0$. Since $xP = 0$ for all $x \in m$, it follows that $\{m\} = \mathrm{Ass}_A(P)$. Since M is a finite A-module, (II,4.3) implies that P has finite A-length; a fortiori, P has finite B-length, so (II,4.3) implies that $\mathrm{Ass}_B(P)$ consists precisely of the maximal ideal of B. Since $\mathrm{Ass}_B(P) \subset \mathrm{Ass}_B(M)$, (3.11) implies that $\mathrm{depth}_B(M) = 0$.

4. Cohen-Macaulay modules and regular local rings.

Definition (4.1). - Let A be a noetherian local ring. A finite A-module M is said to be Cohen-Macaulay if $\mathrm{depth}(M) = \dim(M)$. The ring A is said to be Cohen-Macaulay if it is a Cohen-Macaulay A-module.

Example (4.2). - A noetherian local domain of dimension 1 is Cohen-Macaulay. By Serre's criterion (VII,2.13), a normal noetherian local domain of dimension 2 is Cohen-Macaulay.

Proposition (4.3) (Cohen-Macaulay). - Let A be a noetherian local ring and M a finite A-module. Suppose M is Cohen-Macaulay. Then

(i) M is equidimensional and without embedded primes.

(ii) Let x be an element of the maximal ideal such that $\dim(M/xM) = \dim(M)-1$. Then x is M-regular and M/xM is Cohen-Macaulay.

Proof. By (3.15), $\mathrm{depth}(M) \leq \inf\{\dim(A/p) \mid p \in \mathrm{Ass}(M)\}$ and by (1.1), $\dim(M) = \sup\{\dim(A/p) \mid p \in \mathrm{Ass}(M)\}$; hence, (i) follows from (1.1). Assertion (ii) results from (i) together with (II,3.4), (1.6) and (3.10).

Definition (4.4). - Let B be a noetherian ring, I an ideal of B and A = B/I. Then A is said to be regularly immersed in B if I is generated by a B-regular sequence; more weakly, A is said to be a complete intersection in B if I is generated by $r = \dim(B) - \dim(A)$ elements.

Corollary (4.5). - Let B be a Cohen-Macaulay local ring, I an ideal of B and A = B/I. If A is a complete intersection in B, then A is regularly immersed in B, and A is Cohen-Macaulay.

Definition (4.6). - Let A be a noetherian local ring, m the maximal ideal and $r = \dim(A)$. Then A is said to be regular if m is generated by r elements. Elements of m whose residue classes are linearly independent in m/m^2 are called regular parameters.

Proposition (4.7). - Let A be a noetherian local ring, m the maximal ideal, $k = A/m$ and $r = \dim(A)$. Then:

(i) Elements of m generate if and only if their residue classes generate the k-vector space m/m^2.

(ii) $\dim(A) \leq \dim_k(m/m^2)$, with equality if and only if A is regular.

Proof. Part (i) results immediately from Nakayama's lemma. By (1.9), $\dim(A) \leq s$, the number of elements in a minimal set of generators of m; by (i), $s = \dim_k(m/m^2)$, whence (ii).

Proposition (4.8). - Let A be a noetherian local ring, m the maximal ideal, $k = A/m$ and $x_1,\ldots,x_r \in m$ where $r = \dim(A)$. Then the following conditions are equivalent:

(i) The graded map $k[T_1,\ldots,T_r] \longrightarrow gr_m^*(A)$ defined by $T_i \longmapsto x_i \bmod m^2$ is an isomorphism.

(ii) $x_1,\ldots,x_r$ generate m.

Proof. By (4.7), (i) implies (ii). Assume (ii) and let $S = k[T_1,\dots,T_r]$ and $G = gr^*_m(A)$. Consider the exact sequence $0 \longrightarrow I \longrightarrow S \longrightarrow G \longrightarrow 0$. Now for all positive integers s, $\dim_k(I_s) + \dim_k(G_s) = \dim_k(S_s) = \binom{s+r-1}{r-1}$. Suppose $I \neq 0$. Then for some positive integer h, there exists a nonzero homogeneous element $u \in I_h$ and $I_s \supset uS_{s-h} \cong S_{s-h}$. Therefore, for all $s > h$, $\dim_k(I_s) \geq \dim_k(S_{s-h}) = \binom{s-h+r-1}{r-1}$. Hence, $\dim_k(G_s) \leq f(s) = \binom{s+r-1}{r-1} - \binom{s-h+r-1}{r-1}$. However, $f(s)$ is clearly a polynomial of degree $\leq r-2 = \dim(A) - 2$, contradicting (1.4) (cf.II,4.13); therefore,(i) holds.

Proposition (4.9). - A regular local ring A is a domain.

Proof. Let m be the maximal ideal. By (4.8), $gr^*_m(A)$ is a domain and, by (II,1.15), $\cap\, m^n = 0$. It follows from (II,1.5) that A is a domain.

Proposition (4.10). - Let A be a noetherian local ring, I an ideal of A and $r = \dim(A)$. Then the following conditions are equivalent:

(i) A is regular and I is generated by s regular parameters.

(ii) $B = A/I$ is regular of dimension $r-s$ and I is generated by s elements.

(iii) A is regular and B is regular of dimension $r-s$.

Furthermore, if these conditions hold, I is prime and any s generators are regular parameters.

Proof. Let m be the maximal ideal of A, $m' = m/I$ and $k = A/m$. Then the sequence

$$0 \longrightarrow (m^2 + I)/m^2 \longrightarrow m/m^2 \longrightarrow m'/m'^2 \longrightarrow 0$$

is exact. Assume (i). Then $\dim_k((m^2 + I)/m^2) = s$ and

$\dim_k(m/m^2) = r$, so $\dim_k(m'/m'^2) = r-s$. On the other hand, by (4.7), $\dim_k(m'/m'^2) \geq \dim(B)$ and by (1.6), $\dim(B) \geq r-s$; so, $\dim(B) = r-s$ and B is regular, proving (ii) and (iii).

Assume (ii). Since $\dim_k(m'/m'^2) = r-s$ and $\dim_k(m^2+I)/m^2) \leq s$, $\dim_k(m/m^2) \leq r$. Hence, by (4.7), $\dim_k(m/m^2) = r$ and A is regular. Thus, (iii) holds.

Assume (iii). Then the above exact sequence implies that $\dim_k((m^2 + I)/m^2) = s$. Hence, there exist regular parameters $x_1,\ldots,x_s$ among any set of generators of I. Let I' be the ideal generated by $x_1,\ldots,x_s$. Then by (i)$\Longrightarrow$(ii), A/I' is regular of dimension $r-s$. Thus $I' \subset I$ and by (4.9), they both are primes of coheight $r-s$; hence $I = I'$.

<u>Proposition (4.11)</u>. - Let A be a noetherian local ring, m the maximal ideal and $r = \dim(A)$. Then A is regular if and only if m is generated by an A-regular sequence. Moreover, if $x_1,\ldots,x_r$ are regular parameters of A, then the sequence $(x_1,\ldots x_r)$ is A-regular.

<u>Proof</u>. For $i = 0,\ldots,r$, let I_i be the ideal generated by $x_1,\ldots,x_i$. Then, by (4.10), A/I_i is regular; so, by (4.9), a domain. Hence, x_{i+1} is not a zero-divisor in A/I_i and the sequence $(x_1,\ldots,x_r)$ is A-regular.

Conversely, suppose m is generated by an A-regular sequence $(x_1,\ldots,x_s)$. By (3.15), $s \leq r$ and, by (4.7), $r \leq s$. Hence, $r = s$ and A is regular.

<u>Corollary (4.12)</u>. - A regular local ring is Cohen-Macaulay.

Corollary (4.13). - Let A, B be regular local rings. If A is a quotient of B, then A is regularly immersed (resp. a complete intersection) in B.

5. Homological dimension

Definition (5.1). - Let A be a ring and M an A-module. The projective dimension (resp. injective dimension) of M, denoted $\mathrm{proj.dim}_A(M)$ (resp. $\mathrm{inj.dim}_A(M)$), is defined as the infimum of all integers n such that there exists an exact sequence

$$0 \longrightarrow P_n \longrightarrow \dots \longrightarrow P_0 \longrightarrow M \longrightarrow 0$$

with all P_i projective (resp. an exact sequence

$$0 \longrightarrow M \longrightarrow Q_0 \longrightarrow \dots \longrightarrow Q_n \longrightarrow 0$$

with all Q_i injective).

Proposition (5.2). - Let A be a ring and M an A-module. Then the following conditions are equivalent:

(i) $\mathrm{proj.dim}(M) \leq n$ (resp. $\mathrm{inj.dim}(M) \leq n$).

(ii) $\mathrm{Ext}_A^i(M,N) = 0$ (resp. $\mathrm{Ext}_A^i(N,M) = 0$) for all $i > n$ and all A-modules N.

(ii') $\mathrm{Ext}_A^{n+1}(M,N) = 0$ (resp. $\mathrm{Ext}_A^{n+1}(N,M) = 0$) for all A-modules N.

(iii) In any exact sequence

$$0 \longrightarrow R \longrightarrow P_{n-1} \longrightarrow \dots \longrightarrow P_0 \longrightarrow M \longrightarrow 0$$

with all P_i projective (resp.

$$0 \longrightarrow M \longrightarrow Q_0 \longrightarrow \dots \longrightarrow Q_{n-1} \longrightarrow R \longrightarrow 0$$

with all Q_i injective), R is projective (resp. injective).

Proof. The implications (i) $\Rightarrow$ (ii) and (ii) $\Rightarrow$ (ii') are trivial. To prove the implication (ii') $\Rightarrow$ (iii), note that $\mathrm{Ext}_A^1(R,N) \cong \mathrm{Ext}_A^{n+1}(M,N) = 0$ for all N; hence, R is projective.

Assume (iii) and construct an exact sequence

$$0 \longrightarrow R \longrightarrow P_{n-1} \longrightarrow \dots \longrightarrow P_0 \longrightarrow M \longrightarrow 0$$

with all P_i projective. Then R is projective, so (i) holds. The injectivity statements follow dually.

<u>Lemma (5.3)</u>. - Let A be a ring and N an A-module. Then $\text{inj.dim}(N) \leq n$ if and only if $\text{Ext}_A^{n+1}(A/I,N) = 0$ for all ideals I of A.

<u>Proof</u>. Let

$$0 \longrightarrow N \longrightarrow Q_0 \longrightarrow \dots \longrightarrow Q_{n-1} \longrightarrow R \longrightarrow 0$$

be an exact sequence with all Q_i injective; by (5.2), it suffices to show that R is injective. Now, for all ideals I, $\text{Ext}_A^1(A/I,R) \cong \text{Ext}_A^{n+1}(A/I,N) = 0$; it follows that $\text{Hom}(A,R) \longrightarrow \text{Hom}(I,R)$ is surjective. Consequently, R is injective, ([2],I,3.2).

<u>Definition (5.4)</u>. - Let A be a ring. The <u>global homological dimension</u> of A, denoted $\text{gl.hd}(A)$, is the supremum of the integers $\text{proj.dim}(M)$ as M runs through all A-modules.

<u>Remark (5.5)</u>. - It follows from (5.2) that $\text{gl.hd}(A)$ is the supremum of all integers n for which there exist A-modules M, N such that $\text{Ext}_A^n(M,N) \neq 0$; hence, $\text{gl.hd}(A)$ is the supremum of the integers $\text{inj.dim}(N)$ as N runs through all A-modules.

<u>Proposition (5.6)</u>.- Let A be a ring. Let n be the supremum of the integers $\text{proj.dim}(M)$ as M runs through all finite A-modules. Then $n = \text{gl.hd}(A)$.

<u>Proof</u>. Clearly $n \leq \text{gl.hd}(A)$. On the other hand, for all A-modules N, $\text{Ext}_A^{n+1}(A/I,N) = 0$ for any ideal I; so by (5.3), $\text{inj.dim}(N) \leq n$.

Proposition (5.7). - Let A be a noetherian local ring, k the residue field and M a finite A-module. Let r be an integer satisfying the following conditions:

(i) $\mathrm{Tor}^A_{r+1}(M,k) = 0$

(ii) $\mathrm{Tor}^A_r(M,k) \neq 0$.

Then r is equal to proj.dim(M). Furthermore, if $M \neq 0$ and r = proj.dim(M), then (i) and (ii) hold.

Proof. (ii) implies that proj.dim(M) $\geq r$. On the other hand, consider an exact sequence

$$0 \longrightarrow R \longrightarrow P_{r+1} \longrightarrow \cdots \longrightarrow P_0 \longrightarrow M \longrightarrow 0$$

with all P_i projective of finite type. Since $\mathrm{Tor}^A_1(R,k) \cong \mathrm{Tor}^A_{r+1}(M,k) = 0$, the following lemma implies R is free.

Lemma (5.8). - Let A be a noetherian local ring, k the residue field and R a finite A-module. Then the following conditions are equivalent:

(i) R is free.

(ii) R is projective.

(iii) R is flat.

(iv) $\mathrm{Tor}^A_1(R,k) = 0$.

Proof. The implications (i)$\Longrightarrow$(ii), (ii)$\Longrightarrow$(iii), and (iii)$\Longrightarrow$(iv) are trivial. Assume (iv) and let $x_1,\ldots,x_p$ be elements of R whose images $x_1',\ldots,x_p'$ form a basis of $R\otimes_A k$ over k. Construct the exact sequence

$$A^p \xrightarrow{(x)} R \longrightarrow R'' \longrightarrow 0$$

and consider the exact sequence

$$k^p \xrightarrow{(x')} R\otimes_A k \longrightarrow R''\otimes_A k \longrightarrow 0$$

Since (x') is an isomorphism by construction, $R''/mR'' \cong R''\otimes_A k = 0$ and, hence, by Nakayama's lemma, $R'' = 0$.

Construct the exact sequence

$$0 \longrightarrow R' \longrightarrow A^p \xrightarrow{(x)} R \longrightarrow 0$$

and consider the induced exact sequence

$$\mathrm{Tor}_1^A(R,k) \longrightarrow R'\otimes_A k \longrightarrow k^p \xrightarrow{(x')} R\otimes_A k.$$

Since $\mathrm{Tor}_1^A(R,k) = 0$ by assumption and since (x') is an isomorphism by construction, $R'\otimes_A k = 0$. Hence $R' = 0$ and $A^p \longrightarrow R$ is an isomorphism.

<u>Corollary (5.9)</u>. - Let A be a noetherian local ring and k the residue field. Then $\mathrm{gl.hd}(A) = \mathrm{proj.dim}(k)$.

<u>Proof</u>. The inequality $\mathrm{gl.hd}(A) \geq q = \mathrm{proj.dim}(k)$ is clear. On the other hand, if q is finite, then, for all A-modules M of finite type, $\mathrm{Tor}_{q+1}^A(k,M) = 0$; so, $q \geq \mathrm{proj.dim}(M)$ by (5.7); whence, by (5.6), $q \geq \mathrm{gl.hd}(A)$.

<u>Proposition (5.10)</u>. - Let A be noetherian local ring, m the maximal ideal and M a nonzero, finite A-module. Suppose $x_1 \in m$ is M-regular. Then $\mathrm{proj.dim}(M/x_1M) = \mathrm{proj.dim}(M) + 1$.

<u>Proof</u>. Let $M_1 = M/x_1M$. The exact sequence

$$0 \longrightarrow M \xrightarrow{x_1} M \longrightarrow M_1 \longrightarrow 0$$

yields an exact sequence

$$\mathrm{Tor}_q^A(M,k) \xrightarrow{x_1} \mathrm{Tor}_q^A(M,k) \longrightarrow \mathrm{Tor}_q^A(M_1,k) \longrightarrow \mathrm{Tor}_{q-1}^A(M,k) \xrightarrow{x_1} \mathrm{Tor}_{q-1}^A(M,k)$$

where $k = A/m$. Since $x_1 \in m$, the first and last maps are zero. Take $q = \mathrm{proj.dim}(M) + 1$. Then, by (5.7), $\mathrm{Tor}_q^A(M,k) = 0$ and $\mathrm{Tor}_{q-1}^A(M,k) \neq 0$; hence, $\mathrm{Tor}_q^A(M_1,k) \neq 0$. Now, take $q = \mathrm{proj.dim}(M) + 2$.

Then $\mathrm{Tor}_q^A(M,k) = 0$ and $\mathrm{Tor}_{q-1}^A(M,k) = 0$; hence $\mathrm{Tor}_q^A(M_1,k) = 0$. Therefore, by (5.7), $\mathrm{proj.dim}(M) + 1 = \mathrm{proj.dim}(M_1)$.

Theorem (5.11) (Auslander-Buchsbaum). - Let A be a regular local ring of dimension n. Then $\mathrm{gl.hd}(A) = n$.

Proof. Let $x_1,\ldots,x_n$ be a regular system of parameters of A and k the residue field. Then, by (4.11), $x_1,\ldots,x_n$ is an A-regular sequence and $k = A/(x_1A + \cdots + x_nA)$. So, repeated application of (5.10) yields $\mathrm{proj.dim}(k) = n + \mathrm{proj.dim}(A) = n$; hence, (5.9) yields $n = \mathrm{gl.hd}(A)$.

Lemma (5.12). - Let A be a noetherian local ring and m the maximal ideal. If every element of $m - m^2$ is a zero-divisor, then $m \in \mathrm{Ass}(A)$.

Proof. We may assume $m \neq 0$; whence, by Nakayama's lemma, $m \neq m^2$. By (II,3.5), $m - m^2 \subset \bigcup_{p \in \mathrm{Ass}(A)} p$; hence, $m \subset (\bigcup p) \cup m^2$. By (1.5), $m \subset p$ for some $p \in \mathrm{Ass}(A)$ and, since m is maximal, $m = p$.

Lemma (5.13). - Let A be a noetherian local ring and m the maximal ideal. If $a \in m - m^2$, then m/aA is isomorphic to a direct summand of m/am.

Proof. Let I be an ideal of A such that a and I generate complementary (A/m)-subspaces of m/m^2. Then, by Nakayama's lemma, $I + aA = m$. If $xa \in I$, then its residue class in m/m^2 is zero, so $x \in m$; hence, the natural map $m/aA \cong I/(I \cap aA) \longrightarrow m/am$ is an injection. It is split by the canonical surjection $m/am \longrightarrow m/aA$ and thus m/aA is a direct summand of m/am.

Lemma (5.14). - Let A be a noetherian local ring, m the maximal ideal and M a finite A-module. If $a \in m$ is A-regular and M-regular, then $\text{proj.dim}_{(A/aA)}(M/aM) \leq \text{proj.dim}_A(M)$.

Proof. Clearly we may assume $h = \text{proj.dim}_A(M)$ is finite. If $h = 0$, then by (5.8), M is free and thus M/aM is a free (A/aA)-module; hence, the inequality holds.

Suppose $h \geq 1$. A surjection $E = A^n \longrightarrow M$ yields a commutative diagram

$$\begin{array}{ccccccccc}
0 & \longrightarrow & N & \longrightarrow & E & \longrightarrow & M & \longrightarrow & 0 \\
 & & \downarrow a & & \downarrow a & & \downarrow a & & \\
0 & \longrightarrow & N & \longrightarrow & E & \longrightarrow & M & \longrightarrow & 0 \\
 & & \downarrow & & \downarrow & & \downarrow & & \\
 & & N/aN & \xrightarrow{f} & E/aE & \longrightarrow & M/aM & \longrightarrow & 0 \\
 & & \downarrow & & \downarrow & & \downarrow & & \\
 & & 0 & & 0 & & 0 & &
\end{array}$$

By (5.2), $\text{proj.dim}(N) = h-1$. Furthermore, since a is A-regular, a is E-regular; since a is also M-regular, multiplication by a is injective in all three columns, so by the nine lemma, f is injective. Hence by induction, $\text{proj.dim}_{(A/aA)}(N/aN) \leq h - 1$ and therefore $\text{proj.dim}_{(A/aA)}(M/aM) \leq h$.

Theorem (5.15) (Serre). - If a noetherian local ring A has finite global homological dimension, then it is a regular local ring.

Proof. Let m be the maximal ideal of A, $k = A/m$ and $r = \text{rank}_k(m/m^2)$. If $r = 0$, then by Nakayama's lemma, $m = 0$ and the assertion is trivial.

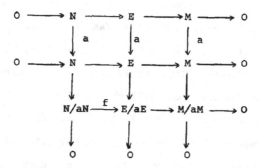

Assume $r \geq 1$. Then k is not projective and thus $q = \text{gl.hd}(A) \geq 1$. Suppose each element of $m - m^2$ is a zero-divisor in A. Then, by (5.12) $m \in \text{Ass}(A)$ and there exists an exact sequence

$$0 \longrightarrow k \xrightarrow{i} A \longrightarrow \text{coker}(i) \longrightarrow 0;$$

it yields an exact sequence

$$0 \longrightarrow \text{Tor}_q^A(k,k) \longrightarrow 0,$$

contradicting (5.7) and (5.9).

Therefore, there is an element $a \in m - m^2$ which is not a zero-divisor. Let $A' = A/aA$ and $m' = m/aA$. Then $\text{rank}_k(m'/m'^2) = r - 1$. By hypothesis, $\text{proj.dim}_A(m)$ is finite; so, by (5.14), $\text{proj.dim}_{A'}(m/am)$ is finite. Since by (5.13), m' is a direct summand of m/am, it follows from (5.2) that $\text{proj.dim}_{A'}(m')$ is finite. It follows from (5.9) that $\text{gl.hd}(A')$ is finite and, by induction, A' is regular of dimension $r - 1$. By (II,3.5) and (1.3), $\dim(A') \leq \dim(A) - 1$ and thus $\dim(A) \geq r$. Hence, by (4.7), $\dim(A) = r$ and A is a regular local ring.

Proposition (5.16). - Let A be a noetherian ring and M an A-module. Then $\text{inj.dim}_A(M) = \sup\{\text{inj.dim}_A(M_m)\}$ where m runs through all prime ideals (resp. maximal ideals) of A. In particular, $\text{gl.hd}(A) = \sup\{\text{gl.hd}(A_m)\}$.

Proof. By (IV,3.2), we have $\text{Ext}^q_{A_m}(A_m/IA_m, M_m) = (\text{Ext}^q_A(A/I,M))_m$ for every prime m and ideal I. Since every ideal of A_m is of the form IA_m, the assertion follows from (II,3.3 and 3.10) and (5.3).

Definition (5.17). - A noetherian ring A is said to be regular if for each prime p of A, the local ring A_p is a regular local ring.

Corollary (5.18). - Let A be a noetherian ring. Then the following conditions are equivalent:

(i) A is regular.

(ii) A_m is a regular local ring for every maximal ideal m of A.

(iii) gl.hd(A) is finite.

Theorem (5.19). - Let A be a regular local ring and M a nonzero, finite A-module. Then

$$\text{depth}(M) + \text{proj.dim}(M) = \dim(A).$$

Proof. If depth(M) = 0, then, by (3.11), the maximal ideal m is in Ass(M). Hence, there exists an exact sequence of the form $0 \longrightarrow k \longrightarrow M \longrightarrow M' \longrightarrow 0$ and it yields an exact sequence

$$\text{Tor}^A_{q+1}(M',k) \longrightarrow \text{Tor}^A_q(k,k) \longrightarrow \text{Tor}^A_q(M,k).$$

Let q = dim(A). By (5.11), $\text{Tor}^A_{q+1}(M',k) = 0$ and, by (5.7), (5.9) and (5.11), $\text{Tor}^A_q(k,k) \neq 0$. Therefore, $\text{Tor}^A_q(M,k) \neq 0$, so proj.dim(M) $\geq$ q; however, q = gl.hd(A), so q = proj.dim(M).

Assume r = depth(M) $\geq$ 1. Then there exists $x \in m$ defining an exact sequence

$$0 \longrightarrow M \xrightarrow{x} M \longrightarrow M_1 \longrightarrow 0.$$

Since $\text{depth}(M_1) = \text{depth}(M) - 1$ by (3.10) and since $\text{proj.dim}(M_1) = \text{proj.dim}(M) + 1$ by (5.10), the assertion follows by induction.

Proposition (5.20). - Let A be a noetherian ring and M a finite A-module. Then proj.dim(M) $\leq$ r if (and only if) $\text{Ext}^{r+1}_A(M,N) = 0$ for all finite A-modules N.

Proof. Consider two exact sequences

$$0 \longrightarrow R \longrightarrow P_{r-1} \longrightarrow \cdots \longrightarrow P_0 \longrightarrow M \longrightarrow 0$$

$$0 \longrightarrow N \longrightarrow P_r \longrightarrow R \longrightarrow 0$$

with all P_i projectives of finite type. Then $\text{Ext}^1_A(R,N) =$

$= \mathrm{Ext}_A^{r+1}(M,N) = 0$; so, $\mathrm{Hom}_A(R,P_r) \longrightarrow \mathrm{Hom}_A(R,R)$ is surjective. Therefore, the second sequence splits and R is projective.

Proposition (5.21). - Let A be a regular local ring, M a finite A-module and r an integer. Then $\mathrm{proj.dim}(M) \leq r$ if (and only if) $\mathrm{Ext}_A^q(M,A) = 0$ for all $q > r$.

Proof. By (5.20), it suffices to show that $\mathrm{Ext}_A^{r+1}(M,N) = 0$ for all finite A-modules N. If $r \geq \mathrm{gl.hd}(A)$, then $\mathrm{Ext}_A^{r+1}(M,N) = 0$ trivially and the proof proceeds by descending induction on r. Consider an exact sequence $0 \longrightarrow P \longrightarrow A^p \longrightarrow N \longrightarrow 0$. It induces an exact sequence

$$\mathrm{Ext}_A^q(M,A^p) \longrightarrow \mathrm{Ext}_A^q(M,N) \longrightarrow \mathrm{Ext}_A^{q+1}(M,P).$$

Thus, for all $q > r$, $\mathrm{Ext}_A^q(M,A^p) = 0$ by hypothesis and $\mathrm{Ext}_A^{q+1}(M,P) = 0$ by induction; hence, $\mathrm{Ext}_A^q(M,N) = 0$.

Corollary (5.22). - Let A be a regular local ring of dimension s and B a quotient of A of dimension s - t. Then B is Cohen-Macaulay if and only if $\mathrm{Ext}_A^q(B,A) = 0$ for all $q > t$.

Proof. By definition and (3.15), B is Cohen-Macaulay if and only if $\dim(B) \leq \mathrm{depth}(B)$. However, by hypothesis, $\dim(B) = s - t$ and, by (5.19) and (3.16), $\mathrm{depth}(B) = s - \mathrm{proj.dim}_A(B)$; the assertion now follows from (5.21).

Chapter IV - Duality Theorems

1. The Yoneda pairing

Theorem (1.1) (Yoneda-Cartier). - Let C and C' be abelian categories and suppose C has enough injectives. Let $T : C \longrightarrow C'$ be an additive, left exact functor. Then, for any two objects F, G in C, there exist pairings

$$R^pT(F) \times \operatorname{Ext}^q(F,G) \longrightarrow R^{p+q}T(G)$$

for all nonnegative integers p and q. These pairings are ∂-functorial; namely, they are functorial in F and G and are compatible with connecting morphisms induced by short exact sequences.

Proof. Choose injective resolutions $0 \longrightarrow F \longrightarrow Q^*(F)$ and $0 \longrightarrow G \longrightarrow Q^*(G)$, and define a complex of abelian groups $\operatorname{Hom}^*(Q^*(F),Q^*(G))$ as follows: Let $\operatorname{Hom}^q(Q^*(F),Q^*(G))$ be the group of all families $u = (u_p)_{p\in Z}$ of morphisms $u_p: Q^p(F) \longrightarrow Q^{p+q}(G)$ (not assumed compatible with the boundary). Define $\partial : \operatorname{Hom}^q(Q^*(F),Q^*(G)) \longrightarrow \operatorname{Hom}^{q+1}(Q^*(F),Q^*(G))$ by $\partial(u) = du + (-1)^q ud$. Then:

(i) $\partial^2 = 0$.

(ii) If $\partial(u) = 0$, then u anti-commutes with the boundary.

(iii) If $v = \partial(u)$, then v is homotopic to zero.

(iv) $H^q(\operatorname{Hom}^*(Q^*(F),Q^*(G)))$ is the group of homotopy classes of morphisms which anti-commute with the boundary

Each $u = (u_p) \in \operatorname{Hom}^q(Q^*(F),Q^*(G))$ induces a morphism $T(u) : TQ^*(F) \longrightarrow TQ^*(G)$ of degree q. If $\partial(u) = 0$, then by (ii), $T(u)$ induces a morphism $H^p(T(u)): R^pT(F) \rightarrow R^{p+q}(G)$ for each p. If $u = \partial(w)$ for some w, then $H^*(T(u)) = 0$ by (iii); hence, $H^*(T(u))$

depends only on the homotopy class of u. Therefore, there exist pairings $R^pT(F) \times H^q(\mathrm{Hom}^*(Q^*(F),Q^*(G)) \longrightarrow R^{p+q}T(G)$; so the following lemma establishes the existence assertion. The ∂-functoriality is straightforward and its proof is omitted.

Lemma (1.2). - Let C be an abelian category, F and G two objects of C and $0 \longrightarrow F \xrightarrow{\varepsilon} Q^*(F)$ and $0 \longrightarrow G \longrightarrow Q^*(G)$ injective resolutions. Then the morphism $\Phi : \mathrm{Hom}^*(Q^*(F),Q^*(G)) \longrightarrow \mathrm{Hom}(F,Q^*(G))$, defined by $\Phi(u) = u \circ \varepsilon$, induces an isomorphism

$$H^q\Phi : H^q(\mathrm{Hom}^*(Q^*(F),Q^*(G))) \xrightarrow{\sim} \mathrm{Ext}^q(F,G)$$

for all $q \geq 0$.

Proof. To construct $H^q(\Phi)^{-1}$, let $a' \in \mathrm{Ext}^q(F,G)$ and choose a representative $a \in \mathrm{Hom}(F,Q^q(G))$ of a'. Since $d \circ a = 0$, a factors through $\ker(d^q)$ and yields a diagram with exact rows

$$\begin{array}{ccccccccc}
0 & \longrightarrow & F & \longrightarrow & Q^0(F) & \longrightarrow & Q^1(F) & \longrightarrow & \cdots \\
 & & \downarrow a & & \downarrow b_0 & & \downarrow b_1 & & \\
0 & \longrightarrow & \ker(d^q) & \longrightarrow & Q^q(G) & \longrightarrow & Q^{q+1}(G) & \longrightarrow & \cdots
\end{array}$$

Since the $Q^q(G)$ are injectives, there exists a morphism $b : Q^*(F) \longrightarrow Q^*(G)$, of degree q, which is unique up to homotopy ([2],V,2.2).

If $a = d \circ s$, then s may be extended to a homotopy (s_p) between b and 0. Therefore, up to homotopy, b depends only on a' If b' is the homotopy class of $((-1)^p b_p)$, then the morphism $a' \longmapsto b'$ is clearly inverse to $H^q(\Phi)$.

Proposition (1.3). - Let C, C' be abelian categories and $S, T : C \longrightarrow C'$ additive functors. Suppose C has enough injectives and S, T are left exact. If $\varepsilon^* : R^*S \longrightarrow R^*T$ is a ∂-morphism of degree r, then, for any two objects F and G, the diagram of Yoneda pairings

$$\begin{array}{ccc}
R^pS(F) \times \mathrm{Ext}^q(F,G) & \longrightarrow & R^{p+q}S(G) \\
\Big\downarrow \varepsilon^p(F) \times \mathrm{id} & & \Big\downarrow \varepsilon^{p+q}(G) \\
R^{p+r}T(F) \times \mathrm{Ext}^q(F,G) & \longrightarrow & R^{p+q+r}T(G)
\end{array}$$

commutes.

__Proof__. For $q = 0$ and all $p \geqslant 0$, the diagram

$$\begin{array}{ccc}
R^pS(F) \times \mathrm{Hom}(F,G) & \longrightarrow & R^pS(G) \\
\Big\downarrow \varepsilon^p(F) \times \mathrm{id} & & \Big\downarrow \varepsilon^p(G) \\
R^{p+r}T(F) \times \mathrm{Hom}(F,G) & \longrightarrow & R^{p+r}T(G)
\end{array}$$

commutes because ε^p is a morphism of functors.

Let $0 \longrightarrow G \longrightarrow Q \longrightarrow G'' \longrightarrow 0$ be an exact sequence with Q injective. Consider the diagram

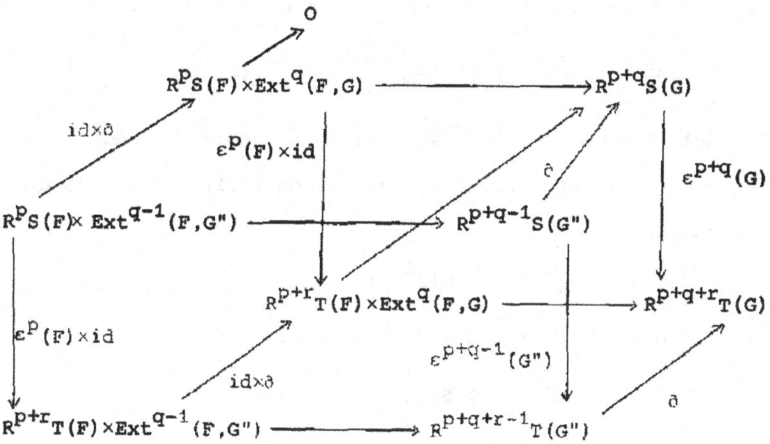

By induction on q, the front face commutes; by (1.1), the horizontal faces commute; and, by hypothesis, the end faces commutes; whence, the assertion.

2. The spectral sequence of a composite functor

Lemma (2.1). - Let C be an abelian category with enough injectives and let $0 \longrightarrow K^0 \longrightarrow K^1 \longrightarrow \dots$ be a complex in C. Then there exists a double complex L^{**}, called a Cartan-Eilenberg resolution of K^*, which gives rise to injective resolutions as follows:

$$0 \longrightarrow K^p \longrightarrow L^{p,0} \longrightarrow L^{p,1} \longrightarrow \dots .$$
$$0 \longrightarrow Z^p(K) \longrightarrow Z^{p,0}(L) \longrightarrow Z^{p,1}(L) \longrightarrow \dots .$$
$$0 \longrightarrow B^p(K) \longrightarrow B^{p,0}(L) \longrightarrow B^{p,1}(L) \longrightarrow \dots .$$
$$0 \longrightarrow H^p(K) \longrightarrow H^{p,0}(L) \longrightarrow H^{p,1}(L) \longrightarrow \dots .$$

Proof. The proof is elementary ([2],XVII,1.2).

Theorem (2.2). - Let C, C' and C'' be abelian categories and suppose C and C' have enough injectives. Let $T : C \longrightarrow C'$ and $S : C' \longrightarrow C''$ be additive functors and suppose S is left exact. Assume that T takes injectives into S-acyclics, i.e., that $(R^qS)(TQ) = 0$ for all $q > 0$ if Q is injective. Then, for any object A of C, there exists a spectral sequence

$$E_2^{p,q} = R^qS(R^pT(A)) \Longrightarrow E^{p+q} = R^{p+q}(S \circ T)(A).$$

Proof. Let $0 \longrightarrow A \longrightarrow Q^*$ be an injective resolution and $0 \longrightarrow T(Q^*) \longrightarrow J^{*,*}$ a Cartan-Eilenberg resolution (2.1). Associated to the double complex

$$S(J^{*,*}) \qquad \begin{array}{ccccccc} 0 & \longrightarrow & S(J^{0,1}) & \longrightarrow & S(J^{1,1}) & \longrightarrow & \dots \\ & & \uparrow & & \uparrow & & \\ 0 & \longrightarrow & S(J^{0,0}) & \longrightarrow & S(J^{1,0}) & \longrightarrow & \dots \\ & & \uparrow & & \uparrow & & \\ 0 & \longrightarrow & ST(Q^0) & \longrightarrow & ST(Q^1) & \longrightarrow & \dots \\ & & \uparrow & & \uparrow & & \\ & & 0 & & 0 & & \end{array}$$

there are two spectral sequences with the same abutment.

In the first spectral sequence ${}_{I}E_1^{p,q} = H_{II}^q(S(J^{p,*})) =$ $= R^qS(T(Q^p))$. However by assumption $R^qS(TQ^p) = 0$ for $q > 0$; so, ${}_{I}E_2^{p,q} = 0$ for $q > 0$. Since S is left exact, ${}_{I}E_2^{p,0} = H_I^p(S\circ T(Q^*)) =$ $= R^p(S\circ T)(A)$. In the second, ${}_{II}E_1^{q,p} = H_I^q(S(J^{*,p})) = SH_I^q(J^{*,p})$; for, $0 \longrightarrow B^{q,p} \longrightarrow Z^{q,p} \longrightarrow H^{q,p} \longrightarrow 0$ splits since $B^{q,p} = B_I^q(J^{*,p})$ is injective. However, $H_I^q(T(Q^*)) = R^qT(A)$ and $0 \longrightarrow H_I^q(T(Q^*)) \longrightarrow H^{q,*}(J^{*,*})$ is an injective resolution, (2.1). Thus ${}_{II}E_2^{q,p} = H_{II}^p(S(H^{q,*}(J^{*,*})) = R^pS(R^qT(A))$, completing the proof.

Lemma (2.3). - Let X be a ringed space. Then the category of O_X-Modules has enough injectives.

Proof. Let F be an O_X-Module and let Q be the O_X-Module defined by $Q(U) = \prod_{x\in U} Q_x$ where Q_x is a fixed injective O_x-module containing F_x and U is any open set of X. Then Q is injective and contains F.

Proposition (2.4). - Let X be a ringed space and F, G two O_X-Modules. Then there exists a spectral sequence

$$H^p(X, \underline{Ext}^q_{O_X}(F,G)) \Longrightarrow Ext^{p+q}_{O_X}(F,G).$$

Proof. $\Gamma(X, \underline{Hom}_{O_X}(F,G)) = Hom_{O_X}(F,G)$; so, the assertion follows from (2.2) in view of (2.3) and the following lemma.

Lemma (2.5). - Let X be a ringed space and F, Q two O_X-Modules. If Q is injective, then $\underline{Hom}_{O_X}(F,Q)$ is flasque.

Proof. Let U be an open subset of X and $f \in \Gamma(U, \underline{Hom}_{O_X}(F,Q))$. Let F_U be the extension of $F|U$ by zero to all of X. Since Q is injective, the map $F_U \longrightarrow Q$ induced by f extends to an element $g \in \Gamma(X, \underline{Hom}_{O_X}(F,Q))$. Then $g|U = f$.

Corollary (2.6). - Let X be a ringed space and E, G two O_X-Modules. If E is locally free of finite type, then $\mathrm{Ext}^p_{O_X}(E,G) = H^p(X, \underline{\mathrm{Hom}}_{O_X}(E,G)) = H^p(X, G \otimes E^\vee)$ where $E^\vee = \underline{\mathrm{Hom}}_{O_X}(E, O_X)$.

Proof. Since E is locally free, the functor $\underline{\mathrm{Hom}}_{O_X}(E,-)$ is exact. It follows that $\underline{\mathrm{Ext}}^q_{O_X}(E,G) = 0$ for all $q > 0$. Hence, the spectral sequence of (2.4) degenerates and $E_2^{p,0} = H^p(X, \underline{\mathrm{Hom}}_{O_X}(E,G))$ is equal to $\mathrm{Ext}^p_{O_X}(E,G)$. The second equality follows from (3.4).

Remark (2.7). - Let $i : X \hookrightarrow P$ be a closed immersion of ringed spaces, E and F two O_X-Modules and G an O_P-Module. Suppose E is locally free. Then it is easily seen that there exist canonical isomorphisms.

(2.7.1) $\mathrm{Hom}_{O_X}(F, \underline{\mathrm{Hom}}_{O_P}(E,G)) \xrightarrow{\sim} \mathrm{Hom}_{O_P}(F \otimes E, G)$

(2.7.2) $\underline{\mathrm{Hom}}_{O_X}(F, \underline{\mathrm{Hom}}_{O_P}(E,G)) \xrightarrow{\sim} \underline{\mathrm{Hom}}_{O_P}(F \otimes E, G)$

Lemma (2.8). - Let $i : X \hookrightarrow P$ be a closed immersion of ringed spaces, Q an injective O_P-Module and E a locally free O_X-Module. Then $J = \underline{\mathrm{Hom}}_{O_P}(E,Q)$ is an injective O_X-Module.

Proof. Let $0 \longrightarrow F' \longrightarrow F$ be an exact sequence of O_X-Modules. Since E is locally free and Q is injective, the sequence $\mathrm{Hom}_{O_P}(F \otimes E, Q) \longrightarrow \mathrm{Hom}_{O_P}(F' \otimes E, Q) \longrightarrow 0$ is exact. Thus by (2.7.1), $\mathrm{Hom}_{O_P}(F,J) \longrightarrow \mathrm{Hom}_{O_X}(F',J) \longrightarrow 0$ is exact.

Proposition (2.9). - Let $X \hookrightarrow P$ be a closed immersion of ringed spaces, E and F two O_X-Modules and G an O_P-Module. Suppose E is locally free of finite type. Then there exist spectral sequences

(2.9.1) $\mathrm{Ext}^p_{O_X}(F, \underline{\mathrm{Ext}}^q_{O_P}(E,G)) \Longrightarrow \mathrm{Ext}^{p+q}_{O_P}(E \otimes F, G)$.

(2.9.2) $\underline{\mathrm{Ext}}^p_{O_X}(F, \underline{\mathrm{Ext}}^q_{O_P}(E,G)) \Longrightarrow \underline{\mathrm{Ext}}^{p+q}_{O_P}(E \otimes F, G)$.

Proof. Apply (2.2) to the functors $\underline{Hom}_{O_P}(E,-)$ and $Hom_{O_X}(F,-)$ (resp. $\underline{Hom}_{O_X}(F,-)$). Then (2.7.1) (resp. 2.7.2) and (2.8) yield (2.9.1) (resp. (2.9.2)).

Remark (2.10) (Leray spectral sequence). - Let $f : X \longrightarrow Y$ be a morphism of ringed spaces. Then the functor f_* is left exact. Furthermore, if Q is an injective O_X-Module, then Q and f_*Q are flasque. By (2.3) and (2.2), there exists a spectral sequence

$$H^p(Y,R^q f_* F) \Longrightarrow H^{p+q}(X,F).$$

3. Complements on $\underline{Ext}^q_{O_X}(F,G)$.

Lemma (3.1). - Let A be a ring, B a flat A-algebra and M,N two A-modules. Suppose M has a presentation $E_q \longrightarrow \dots \longrightarrow E_0 \longrightarrow M \longrightarrow 0$ where the E_i are finite, free A-modules. Then the canonical B-homomorphisms

$$\mathrm{Ext}^r_A(M,N)\otimes_A B \longrightarrow \mathrm{Ext}^r_B(M\otimes_A B, N\otimes_A B)$$

are isomorphisms for $0 \leq r < q$.

Proof. Consider the commutative diagram with exact rows,

$$\begin{array}{ccccccc}
0 \longrightarrow & \mathrm{Hom}_A(M,N)\otimes_A B & \longrightarrow & \mathrm{Hom}_A(E_0,N)\otimes_A B & \longrightarrow & \mathrm{Hom}_A(E_1,N)\otimes_A B \\
 & \downarrow f & & \downarrow g & & \downarrow h \\
0 \longrightarrow & \mathrm{Hom}_B(M\otimes_A B,N\otimes_A B) & \longrightarrow & \mathrm{Hom}_B(E_0\otimes_A B,N\otimes_A B) & \longrightarrow & \mathrm{Hom}_B(E_1\otimes_A B,N\otimes_A B),
\end{array}$$

since g and h are clearly isomorphisms, f is an isomorphism.

Let $M' = \ker(E_0 \longrightarrow M)$ and consider the commutative diagram with exact rows,

$$\begin{array}{ccccccc}
\mathrm{Ext}^{q-1}_A(E_0,N)\otimes_A B & \longrightarrow & \mathrm{Ext}^{q-1}_A(M',N)\otimes_A B & \longrightarrow & \mathrm{Ext}^q_A(M,N)\otimes_A B & \longrightarrow 0 \\
\downarrow & & \downarrow & & \downarrow & \\
\mathrm{Ext}^{q-1}_B(E_0\otimes_A B,N\otimes_A B) & \longrightarrow & \mathrm{Ext}^{q-1}_B(M'\otimes_A B,N\otimes_A B) & \longrightarrow & \mathrm{Ext}^q_B(M\otimes_A B,N\otimes_A B) & \longrightarrow 0.
\end{array}$$

Thus, the assertion follows by induction.

Proposition (3.2). - Let X be a locally noetherian scheme and F, G two coherent O_X-Modules. Then, for all q:

(i) $\underline{\mathrm{Ext}}^q_{O_X}(F,G)$ is coherent

(ii) If $X = \mathrm{Spec}(A)$, $F = \tilde{M}$ and $G = \tilde{N}$, then $\underline{\mathrm{Ext}}^q_{O_X}(F,G) = \mathrm{Ext}^q_A(M,N)^\sim$.

(iii) For any point $x \in X$, $\underline{\mathrm{Ext}}^q_{O_X}(F,G)_x = \mathrm{Ext}^q_{O_x}(F_x,G_x)$.

(iv) If X is a scheme projective over a noetherian ring k, then $\mathrm{Ext}^q_{O_X}(F,G)$ is a finite k-module.

Proof. Clearly, (i) follows from (ii); (ii) from (3.1); (iii) from (ii) and (3.1). Furthermore, (iv) follows from (i), (2.4) and part (i) of the following proposition.

Proposition (3.3) (Serre; [7] III, 2.2.2). - Let k be a noetherian ring, X a projective k-scheme and F a coherent O_X-Module. Then:

(i) The k-modules $H^q(X,F)$ are of finite type.

(ii) There exists an integer m_0 such that for all $m \geq m_0$ and all $q > 0$, $H^q(X,F(m)) = 0$.

(iii) There exists an integer m_0 such that for all $m \geq m_0$, $H^0(X,F(m))$ generates $F(m)$.

Proposition (3.4). - Let X be a ringed space, E, F, G three O_X-Modules and $E^\vee = \underline{\mathrm{Hom}}_{O_X}(E,O_X)$. Suppose E is locally free of finite rank. Then the canonical homomorphisms

$$\underline{\mathrm{Ext}}^q_{O_X}(F,G) \otimes_{O_X} E^\vee \longrightarrow \underline{\mathrm{Ext}}^q_{O_X}(E \otimes_{O_X} F,G)$$

are isomorphisms for all $q \geq 0$.

Proof. The map $\underline{\mathrm{Ext}}^q_{O_X}(F,G) \otimes E^\vee \longrightarrow \underline{\mathrm{Ext}}^q_{O_X}(E \otimes F,G)$ is clearly an isomorphism for $E = O_X$ and hence also for $E = O_X^n$. Since E is locally free and the map is globally defined, it is therefore an isomorphism.

4. Serre duality

Proposition (4.1) (Serre, [7] III, 2.1.12). - Let k be a ring $P = \mathbb{P}^n_k (= \mathrm{Proj}(k[T_0,\ldots,T_n]))$. Then

(i) $H^q(P,O_P(r)) = 0$ for all r and all $q \neq 0,n$.

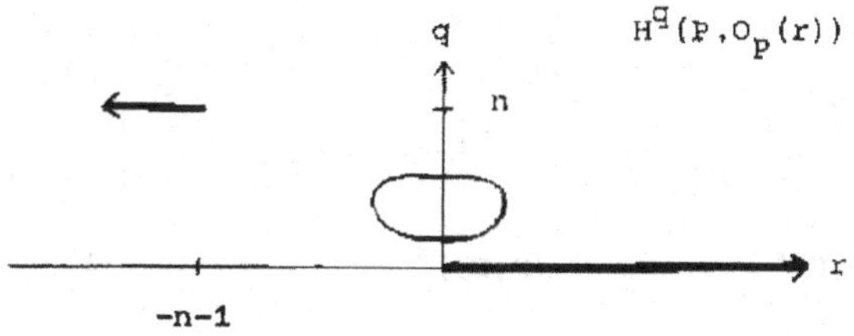

(ii) The canonical homomorphism $k[T_0,\ldots,T_n] \longrightarrow \bigoplus_q H^0(P,O_P(q))$ is bijective.

(iii) $H^n(P,O_P(-m-n-1))$ is the free module on symbols $\xi_{p_0,\ldots,p_n}$ where the p_i are nonnegative integers and $\Sigma p_i = m$. Furthermore, $T_i \xi_{p_0,\ldots,p_n} = \xi_{p_0,\ldots,p_i-1,\ldots,p_n}$ if $p_i > 0$ or $= 0$ if $p_i = 0$.

Theorem (4.2). - Let k be a field, $P = \mathbb{P}^n_k$ and $\omega_P = O_P(-n-1)$. Then Yoneda pairing $H^r(P,F) \times \mathrm{Ext}^{n-r}_{O_P}(F,\omega_P) \longrightarrow H^n(P,\omega_P)$ is nonsingular; that is, there is an isomorphism $\eta : H^n(P,\omega_P) \xrightarrow{\sim} k$ and the induced map $y_r(F)\colon \mathrm{Ext}^{n-r}_{O_P}(F,\omega_P) \longrightarrow H^r(P,F)^*$ is an isomorphism of ∂-functors in F.

Proof. With $F = O_P(-m-n-1)$ and $r = n$, the pairing becomes

$$H^n(P,O_P(-m-m-1)) \times H^0(P,O_P(m)) \longrightarrow H^n(P,O_P(-n-1))$$

$$(\xi_{p_0,\ldots,p_n}, T_0^{q_0}\ldots T_n^{q_n}) \longrightarrow T_0^{q_0}\ldots T_n^{q_n}\xi_{p_0,\ldots,p_n}.$$

However, $T_0^{q_0}\cdots T_n^{q_n}\xi_{p_0,\dots,p_n} = \xi_{0,\dots,0}$ if $q_i = p_i$ for all i and $= 0$ otherwise. Hence, the bases $\{T_0^{q_0}\cdots T_n^{q_0} \mid \Sigma q_i = m\}$ and $\{\xi_{p_0,\dots,p_n} \mid \Sigma p_i = m\}$ are dual and the pairing is nonsingular in this case.

In general, by (3.3), there is a presentation

$$E_1 \longrightarrow E_0 \longrightarrow F \longrightarrow 0$$

where the E_i are of the form $O_P(-m)^q$ for suitable integers $m, q > 0$. Consider the diagram

$$\begin{array}{ccccccc}
0 \longrightarrow & \mathrm{Hom}_{O_P}(F,\omega_P) & \longrightarrow & \mathrm{Hom}_{O_P}(E_0,\omega_P) & \longrightarrow & \mathrm{Hom}_{O_P}(E_1,\omega_P) \\
& \downarrow{\scriptstyle y_n(F)} & & \downarrow{\scriptstyle y_n(E_0)} & & \downarrow{\scriptstyle y_n(E_1)} \\
0 \longrightarrow & H^n(P,F)^* & \longrightarrow & H^n(P,E_0)^* & \longrightarrow & H^n(P,E_1)^*
\end{array}$$

where the y_n arise from the isomorphism $\eta : H^n(P,\omega_P) \xrightarrow{\sim} k$ defined by $\eta(a\xi_{0,\dots,0}) = a$. It results from the preceding paragraph that the $y_n(E_i)$ are isomorphisms. The diagram is commutative by the functoriality of the Yoneda pairing and its bottom row is exact by the right exactness of $H^n(P,-)$. Hence, $y_n(F)$ is an isomorphism.

Consider an exact sequence of the form $0 \longrightarrow G \longrightarrow E \longrightarrow F \longrightarrow 0$ where $E = O_P(-m)^q$ for suitable integers $m, q > 0$. The diagram

$$\begin{array}{ccccccc}
\mathrm{Ext}^{n-r-1}_{O_P}(E,\omega_P) & \longrightarrow & \mathrm{Ext}^{n-r-1}_{O_P}(G,\omega_P) & \longrightarrow & \mathrm{Ext}^{n-r}_{O_P}(F,\omega_P) & \longrightarrow & \mathrm{Ext}^{n-r}_{O_P}(E,\omega_P) \\
\downarrow{\scriptstyle y_{r+1}(E)} & & \downarrow{\scriptstyle y_{r+1}(G)} & & \downarrow{\scriptstyle y_r(F)} & & \downarrow{\scriptstyle y_r(E)} \\
H^{r+1}(P,E)^* & \longrightarrow & H^{r+1}(P,G)^* & \longrightarrow & H^r(P,F)^* & \longrightarrow & H^r(P,E)^*
\end{array}$$

is commutative by the ∂-functoriality of the Yoneda pairing. If $r < n$, then $y_{r+1}(E)$ and $y_{r+1}(G)$ are isomorphisms by descending induction and $H^r(P,E) = 0$ by (4.1). Finally, it follows from (2.6) and (4.1) that $\mathrm{Ext}^{n-r}_{O_P}(E,\omega_P) = H^{n-r}(P,\omega_P(m))^q = 0$. The proof of Serre duality is now complete.

5. Grothendieck duality

<u>Lemma (5.1)</u>. - Let k be a field, P a regular k-scheme of pure dimension n and X a closed subscheme of P, ω_P an invertible sheaf on P. Suppose X has pure dimension r (i.e., every irreducible component has dimension r). Then $\underline{\mathrm{Ext}}^q_{O_P}(O_X,\omega_P) = 0$ for $q < n-r$.

<u>Proof</u>. By (III,3.13), $\underline{\mathrm{Ext}}^q_{O_P}(O_X,\omega_P) = 0$ for $q < d =$ $= \inf_{x\in X}\{\mathrm{depth}(\omega_{P,x})\}$. Since ω_P is invertible, $\omega_{P,x} = O_{P,x}$ and, since $O_{P,x}$ is regular, $\mathrm{depth}(O_{P,x}) = \dim(O_{P,x})$ by (III,4.12). Therefore, $d = n-r$ and the proof is complete.

<u>Lemma (5.2)</u>. - Under the conditions of (5.1), there exists a ∂-morphism $\varepsilon^* : \mathrm{Ext}^*_{O_X}(-,\omega_X) \longrightarrow \mathrm{Ext}^*_{O_P}(-,\omega_P)$ of degree r where $\omega_X = \underline{\mathrm{Ext}}^{n-r}_{O_P}(O_X,\omega_P)$.

<u>Proof</u>. Let F be a coherent O_X-Module and consider the spectral sequence (2.9.1)

$$E_2^{t,s} = \mathrm{Ext}^t_{O_X}(F,\underline{\mathrm{Ext}}^s_{O_P}(O_X,\omega_P)) \Longrightarrow \mathrm{Ext}^{s+t}_{O_P}(F,\omega_P).$$

By (5.1), $E_2^{t,s} = 0$ for $s < n-r$.

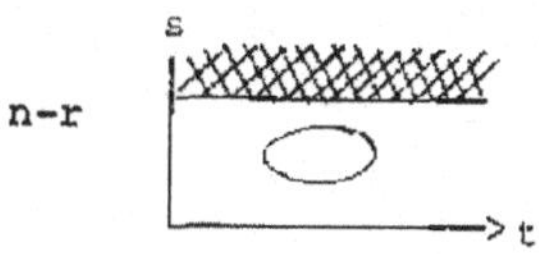

Let $\varepsilon^{r-p}(F) : \mathrm{Ext}^{r-p}_{O_X}(F,\omega_X) \longrightarrow \mathrm{Ext}^{n-p}_{O_P}(F,\omega_P)$ be the edge homomorphisms.

Given an exact sequence of O_X-Modules $0 \longrightarrow F' \longrightarrow F \longrightarrow F'' \longrightarrow 0$, we deduce an exact sequence of double complexes

$$0 \longrightarrow \mathrm{Hom}(F'', J^{*,*}) \longrightarrow \mathrm{Hom}(F, J^{*,*}) \longrightarrow \mathrm{Hom}(F', J^{*,*}) \longrightarrow 0,$$ where $J^{*,*}$ is as in (2.2), and thence a cohomology triangle of spectral sequences

$$\begin{array}{ccc} \mathrm{Ext}^*_{O_X}(F'', \underline{\mathrm{Ext}}^*_{O_P}(O_X, \omega_P)) & \longrightarrow & \mathrm{Ext}^*_{O_X}(F, \underline{\mathrm{Ext}}^*_{O_P}(O_X, \omega_P)) \\ {\scriptstyle\partial}\nwarrow & & \swarrow \\ & \mathrm{Ext}^*_{O_X}(F', \underline{\mathrm{Ext}}^*_{O_P}(O_X, \omega_P)). & \end{array}$$

It follows that ε^* is a map of ∂-functors.

Lemma (5.3). - Under the conditions of (5.2), if F is a coherent O_X-Module, then the following diagram commutes:

$$\begin{array}{ccc} H^p(X,F) \times \mathrm{Ext}^{r-p}_{O_X}(F,\omega_X) & \longrightarrow & H^r(X,\omega_X) \\ \Big\downarrow {\scriptstyle \varepsilon^{r-p}(F) \times \mathrm{id}} & & \Big\downarrow {\scriptstyle i} \\ H^p(P,F) \times \mathrm{Ext}^{n-p}_{O_P}(F,\omega_P) & \longrightarrow & H^n(P,\omega_P) \end{array}$$

where i is the map induced by $\varepsilon_0(\omega_X)(\mathrm{id}_{\omega_X}) \in \mathrm{Ext}^{n-r}_{O_P}(\omega_X,\omega_P)$ via the Yoneda pairing.

Proof. Given $f \in \mathrm{Ext}^{r-p}_{O_X}(F,\omega_X)$, consider the diagram

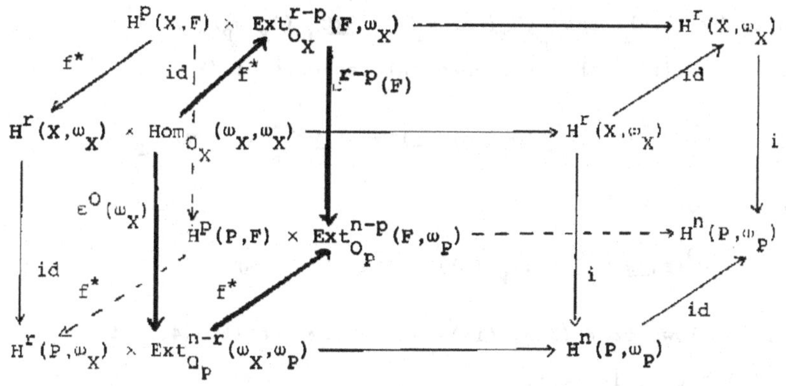

where the horizontal maps f^* are induced by f via Yoneda pairing and the rows are Yoneda pairings. If $a \in H^p(X,F)$, then $\langle a,f\rangle =$ $= \langle f^*(a), id_{\omega_X}\rangle$ and $\langle a, f^*(\varepsilon_0(\omega_X)(id_{\omega_X}))\rangle = \langle f^*(a), \varepsilon_0(\omega_X)(id_{\omega_X})\rangle$ by (1.1). By (5.2) and (1.3), the darkened square commutes; whence, the assertion.

<u>Theorem (5.4)</u>. - Let k be a field, $P = \mathbb{P}^n_k$, and $\omega_P = O_P(-n-1)$. Let X be a closed subscheme of P of pure dimension r and F a coherent O_X-Module. Then for every integer $s \leq r$, the following conditions are equivalent:

(i) Let $\eta_P\colon H^n(P,\omega_P) \longrightarrow k$ be a k-linear isomorphism and $\eta_X = \eta_P \circ i$. Then the corresponding map $\mathrm{Ext}^{r-p}_{O_X}(F,\omega_X) \longrightarrow H^p(X,F)^*$ is an isomorphism for $r-s \leq p \leq r$.

(ii) $H^p(X,O_X(-m)) = 0$ for large m and for $r-s \leq p < r$.

(iii) $\underline{\mathrm{Ext}}^{n-p}_{O_P}(O_X,\omega_P) = 0$ for $r-s \leq p < r$.

<u>Proof</u>. Assume (i). Then $H^p(X,O_X(-m)) = 0$ if (and only if) $\mathrm{Ext}^{r-p}_{O_X}(O_X(-m),\omega_X) = 0$. However, by (2.6), $\mathrm{Ext}^{r-p}_{O_X}(O_X(-m),\omega_X) =$ $= H^{r-p}(X,\omega_X(m))$ and by (3.3,(ii)), $H^{r-p}(X,\omega_X(m)) = 0$ for large m. Thus (ii) holds.

Since, by (3.4), $\underline{\mathrm{Ext}}^q_{O_P}(O_X(-m),\omega_P) = \underline{\mathrm{Ext}}^q_{O_P}(O_X,\omega_P)(m)$, it follows from (3 3,(ii)) that the spectral sequence of (2.4).

$$H^{n-p-q}(P,\underline{\mathrm{Ext}}^q_{O_P}(O_X(-m),\omega_P)) \Longrightarrow \underline{\mathrm{Ext}}^{n-p}_{O_P}(O_X(-m),\omega_P)$$

degenerates and yields

$$H^0(P,\underline{\mathrm{Ext}}^{n-p}_{O_P}(O_X,\omega_P)(m)) = \mathrm{Ext}^{n-p}_{O_P}(O_X(-m),\omega_P).$$

It therefore follows from (3.3,(iii)) and Serre duality (4.2) that (ii) and (iii) are equivalent.

Assume (iii). Then in the spectral sequence (2.9.1)

$$E_2^{t,q} = \mathrm{Ext}^{t,q}_{O_X}(F,\underline{\mathrm{Ext}}^q_{O_P}(O_X,\omega_P)) \Longrightarrow \mathrm{Ext}^{n-p}_{O_P}(F,\omega_P)$$

where $t = n-p-q$, we have $E_2^{t,q} = 0$ for $n-r < q \leq n-r+s$ and for $q < n-r$ by (5.1).

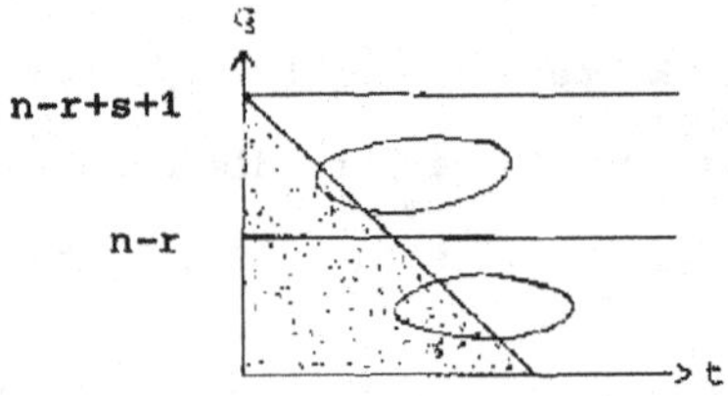

Therefore, for $t = r-p < s+1$, the edge homomorphism $\varepsilon^{r-p}(F)$ is an isomorphism. However, by (5.3), the diagram

$$\begin{array}{ccc} \mathrm{Ext}^{r-p}_{O_X}(F,\omega_X) & \longrightarrow & H^p(X,F)^* \\ \downarrow {\scriptstyle \varepsilon^{r-p}(F)} & & \downarrow {\scriptstyle \mathrm{id}} \\ \mathrm{Ext}^{n-p}_{O_P}(F,\omega_P) & \longrightarrow & H^p(P,F)^* \end{array}$$

is commutative. Hence, (i) results from Serre duality (4.2).

Corollary (5.5). - Under the conditions of (5.4), the map

$$\mathrm{Hom}_{O_X}(F,\omega_X) \longrightarrow H^r(X,F)^*$$

is always an isomorphism.

Corollary (5.6). - Under the conditions of (5.4), the map

$$\mathrm{Ext}^{r-p}_{O_X}(F,\omega_X) \longrightarrow H^p(X,F)^*$$

is an isomorphism for all p if and only if X is Cohen-Macaulay.

Proof. The assertion results immediately from (III,5.22), (5.4) and (3.2).

Chapter V - Flat Morphisms

1. Faithful flatness

Let C, C' be categories and $T : C \longrightarrow C'$ a functor. Then T is said to be <u>faithful</u> if, for all $M, N \in C$, the canonical map $\mathrm{Hom}(M,N) \longrightarrow \mathrm{Hom}(TM,TN)$ is injective. If C, C' are additive and T is additive, then clearly T is faithful if and only if, for all maps $u : M \longrightarrow N$, $T(u) = 0$ implies $u = 0$.

<u>Proposition (1.1)</u>. - If C, C' are abelian categories and $T : C \longrightarrow C'$ is an additive functor, then the following conditions are equivalent:

(i) T is exact and faithful.

(ii) T is exact and, for all $N \in C$, $TN = 0$ implies $N = 0$.

(iii) A sequence $N' \longrightarrow N \longrightarrow N''$ in C is exact if and only if $TN' \longrightarrow TN \longrightarrow TN''$ is exact.

<u>Proof</u>. Assume (i). Then $TN = 0$ implies $T(id_N) = 0$; hence, $id_N = 0$ and $N = 0$; thus, (ii) holds. In (iii), suppose $TN' \xrightarrow{Tu} TN \xrightarrow{Tv} TN''$ is exact. Then $Tvu = TvTu = 0$; so, $vu = 0$ by (i). Let $I = \mathrm{im}(u)$, $K = \ker(v)$, $i : I \longrightarrow K$ and $K' = \mathrm{coker}(i)$. Since T is exact, it follows that $TK' = 0$; so, $K' = 0$ by (i)$\Longrightarrow$(ii). Thus, $N' \longrightarrow N \longrightarrow N''$ is exact and (iii) holds.

Let $u : N' \longrightarrow N$ be such that $Tu = 0$. If (iii) holds, consider the map $v : N \longrightarrow \mathrm{coker}(u)$. Tv is an isomorphism, so v is an isomorphism and $u = 0$; hence, (i) holds. If (ii) holds, consider $I = \mathrm{im}(u)$. $T(I) = 0$, so $I = 0$ and $u = 0$; hence, (i) holds.

Corollary (1.2). - Under the conditions of (1.1), suppose there exists a family $\{N_\alpha\}$ of objects of C such that, for each nonzero object N of C, there exist exact sequences $0 \longrightarrow N' \longrightarrow N$ and $N' \longrightarrow N_\beta \longrightarrow 0$ for suitable N' and N_β. Then T is exact and faithful if and only if T is exact and $TN_\beta \neq 0$ for all N_β.

Definition (1.3). - Let A be a ring. An A-module M is said to be faithfully flat over A if the functor $M \otimes_A -$ is exact and faithful.

Proposition (1.4). - Let A be a ring and M an A-module. Then the following conditions are equivalent:

(i) M is faithfully flat.

(ii) M is flat and, if N is an A-module such that $M \otimes_A N = 0$, then $N = 0$.

(iii) M is flat and, for all maximal ideals m, $M \otimes_A (A/m) \neq 0$.

(iv) A sequence of A-modules $N' \longrightarrow N \longrightarrow N''$ is exact if and only if $M \otimes_A N' \longrightarrow M \otimes_A N \longrightarrow M \otimes_A N''$ is exact.

Proof. Let N be a nonzero A-module. Then there exists an injection of the form $0 \longrightarrow A/I \longrightarrow N$ where I is a proper ideal of A; further, there exists a surjection $A/I \longrightarrow A/m \longrightarrow 0$ where m is a maximal ideal Therefore, the equivalence follows from (1.1) and (1.2).

Proposition (1.5). - Let $A \longrightarrow B$ be a ring homomorphism, M, N two A-modules and P a B-module. Then:

(i) If M and N are flat (resp. faithfully flat) over A, then $M \otimes_A N$ is flat (resp. faithfully flat) over A.

(ii) If M is flat (resp. faithfully flat) over A, then $M \otimes_A B$ is flat (resp. faithfully flat) over B.

(iii) If B is flat (resp. faithfully flat) over A and P is flat (resp. faithfully flat) over B, then P is flat (resp. faithfully flat) over A.

(iv) If B is faithfully flat over A and $M\otimes_A B$ is flat (resp. faithfully flat) over B, then M is flat (resp. faithfully flat) over A.

Proof. The assertions result easily from the following formulas, functorial in R: $(M\otimes_A N)\otimes_A R = M\otimes_A(N\otimes_A R)$; $(M\otimes_A B)\otimes_B R = M\otimes_A R$; $P\otimes_B R = P\otimes_B(B\otimes_A R)$; and $(M\otimes_A R)\otimes_A B = (M\otimes_A B)\otimes_B(R\otimes_A B)$.

Proposition (1.6). - Let $\varphi : A \longrightarrow B$ be a local homomorphism of rings and M a B-module of finite type. Then M is faithfully flat over A if (and only if) M is flat over A and $M \neq 0$. In particular, B is faithfully flat over A if (and only if) B is flat over A.

Proof. Let m (resp. n) be the maximal ideal of A (resp. B). By (1.4), it suffices to show that $M\otimes_A(A/m) \neq 0$. However, if $M\otimes_A(A/m) = 0$, then $nM = M$; so, by Nakayama's lemma, $M = 0$.

Lemma (1.7). - Let A be a ring and M an A-module. Then M is flat if (and only if) $\mathrm{Tor}_1^A(M,A/I) = 0$ for all ideals I.

Proof. If N is an A-module generated by r elements, there exists a submodule N' of N generated by $r-1$ elements such that $N/N' = A/I$ for some ideal I. The sequence

$$\mathrm{Tor}_1^A(M,N') \longrightarrow \mathrm{Tor}_1^A(M,N) \longrightarrow \mathrm{Tor}_1^A(M,N/N')$$

is then exact. It follows by induction on r that $\mathrm{Tor}_1^A(M,N) = 0$ for all A-modules N of finite type. Finally, since any A-module is the inductive limit of its submodules of finite type and since the functor $\mathrm{Tor}_1^A(M,-)$ commutes with inductive limits, $\mathrm{Tor}_1^A(M,N) = 0$ for all A-modules N.

<u>Lemma (1.8)</u>. - Let A be a ring and M an A-module. For any ideal I of A, $\mathrm{Tor}_1^A(M,A/I) = 0$ if and only if the canonical surjection $I\otimes_A M \longrightarrow IM$ is bijective.

<u>Proof.</u> The assertion results immediately from the exact sequence

$$0 \longrightarrow \mathrm{Tor}_1^A(M,A/I) \longrightarrow I\otimes_A M \longrightarrow M.$$

<u>Theorem (1.9)</u>. - Let $\varphi : A \longrightarrow B$ be a ring homomorphism. Then the following conditions are equivalent:

(i) B is faithfully flat over A.

(ii) φ is injective and $B/\varphi(A)$ is flat over A.

(iii) B is flat over A and, for any A-module M, $\mathrm{id}_M\otimes\varphi : M \longrightarrow M\otimes_A B$ is injective.

(iv) For any ideal I of A, the natural map $I\otimes_A B \longrightarrow IB$ is bijective and $\varphi^{-1}(IB) = I$.

<u>Proof</u>. Assume (i) and consider the sequence $0 \longrightarrow N \longrightarrow M \xrightarrow{u} M\otimes_A B$ where $N = \ker(u)$. Then the sequence $0 \longrightarrow N\otimes_A B \longrightarrow M\otimes_A B \xrightarrow{u\otimes \mathrm{id}_B} M\otimes_A B\otimes_A B$ is exact and $u\otimes\mathrm{id}_B$ has a left inverse induced by the canonical map $B\otimes_A B \longrightarrow B$; hence, $N\otimes_A B = 0$. Thus $N = 0$ and (iii) holds.

If the sequence $0 \longrightarrow A \xrightarrow{\varphi} B \longrightarrow B/\varphi(A) \longrightarrow 0$ is exact, it yields an exact sequence

$$0 \longrightarrow \mathrm{Tor}_1^A(M,B) \longrightarrow \mathrm{Tor}_1^A(M,B/\varphi(A)) \longrightarrow M \longrightarrow M\otimes_A B$$

for all A-modules M. It follows that (ii) and (iii) are equivalent.

Assume (iii). By (1.8), $I\otimes_A B \longrightarrow IB$ is bijective; so, $0 \longrightarrow A/I \longrightarrow B/IB = (A/I)\otimes B$ is exact and it follows that $\varphi^{-1}(IB) = I$. Thus, (iv) holds.

Finally, assume (iv); by (1.8), $\mathrm{Tor}_1^A(B,A/I) = 0$ and, thus, by (1.7), B is flat. If m is a maximal ideal of A, then $\varphi^{-1}(mB) = m$ implies $mB \neq B$; so $0 \neq B/mB = B\otimes_A(A/m)$. By (1.4), B is faithfully flat over A.

<u>Proposition (1.10)</u>. - Let A be a noetherian ring and q an ideal of A. Then $\hat{A} = \varprojlim A/q^r$ is a flat A-module. Furthermore, Â is faithfully A-flat if and only if $q \subset \mathrm{rad}(A)$.

<u>Proof</u>. The functor $M \longmapsto \hat{A}\otimes_A M$ is exact for finite A-modules M by (II,1.17 and 1.18). If there were an injection $N' \longrightarrow N$ such that $N'\otimes_A\hat{A} \longrightarrow N\otimes_A\hat{A}$ is not injective, then there would be a sub-injection $M' \longrightarrow M$ of finite submodules such that $M'\otimes_A\hat{A} \longrightarrow M\otimes_A\hat{A}$ is not injective; hence Â is flat.

If m is a maximal ideal of A, then, by (II,1.18), $\hat{A}\otimes_A A/m = (A/m)\hat{} = \varprojlim A/(q^r + m)$; so, $\hat{A}\otimes_A A/m \neq 0$ if and only if $q \subset m$. Therefore the last assertion follows from (1.4).

2. Flat morphisms

<u>Definition (2.1)</u>. - Let $f : X \longrightarrow Y$ be a morphism of local-ringed spaces and F an O_X-Module. Then F is said to be <u>flat over</u> Y <u>at</u> $x \in X$ if F_x is $O_{f(x)}$-flat, to be <u>flat over</u> $y \in Y$ if F is flat over Y at every $x \in f^{-1}(y)$, to be <u>flat over</u> Y if F is flat over every $y \in Y$ and to be <u>faithfully flat</u> over Y if F is flat over Y and $F\otimes k(y) \neq 0$ for every $y \in Y$.

<u>Proposition (2.2)</u>. - Let $f : X \longrightarrow Y$ be a morphism of affine schemes and F a quasi-coherent O_X-Module. Then F is flat (resp. faithfully flat) over Y if and only if $M = \Gamma(X,F)$ is flat (resp. faithfully flat) over $A = \Gamma(Y,O_Y)$.

Proof. Given a sequence $0 \longrightarrow N' \longrightarrow N$ of A-modules, the sequence $0 \longrightarrow M\otimes_A N' \longrightarrow M\otimes_A N$ is exact if (and only if) the sequence $0 \longrightarrow \tilde{M}\otimes_{O_Y} \tilde{N}' \longrightarrow \tilde{M}\otimes_{O_Y} \tilde{N}$ is exact. Thus, if $F = \tilde{M}$ is flat, then M is is flat; further, if F is faithfully flat, then M is faithfully flat by (1.4). The converse results from the following lemma.

Lemma (2.3). - Let A be a ring, B an A-algebra and S (resp. T) a multiplicative set in A (resp. B) such that S maps into T. If a B-module M is flat over A, then $T^{-1}M$ is flat over $S^{-1}A$.

Proof. If N is an $(S^{-1}A)$-module, then $T^{-1}M\otimes_{S^{-1}A} N \cong$ $= T^{-1}(M\otimes_A N)$; hence, the functor $T^{-1}M\otimes_{S^{-1}A}-$ is the composite of the exact functors $M\otimes_A -$ and $T^{-1}-$.

Proposition (2.4). - Let $f : X \longrightarrow Y$ be a morphism of schemes and F a quasi-coherent O_X-Module of finite type. Then F is faithfully flat over Y if (and only if) F is flat over Y and $f(\mathrm{Supp}(F)) = Y$.

Proof. It suffices to show that $F\otimes_{O_Y} O_y \neq 0$ if and only if $F\otimes_{O_Y} k(y) \neq 0$. However, if $F\otimes k(y) \neq 0$, then, clearly, $F\otimes O_y \neq 0$; conversely, if $F\otimes O_y \neq 0$, then there exists a point $x \in X$ such that $f(x) = y$ and $F_x \neq 0$. Therefore $m_y F_x \subset m_x F_x \neq F_x$ by Nakayama's lemma; so, $F\otimes_{O_Y} k(y) \neq 0$.

Definition (2.5). - A morphism of schemes $f : X \longrightarrow Y$ is said to be quasi-flat if there exists a quasi-coherent O_X-Module F of finite type which is flat over Y and whose support is X. Further f is said to be quasi-faithfully flat if f is quasi-flat and surjective. Finally, f is said to be flat (resp. faithfully flat) if O_X is flat over Y (resp. O_X is flat over Y and f is surjective).

Corollary (2.6). - Let f be a quasi-flat morphism of schemes. Let $x \in X$ and $y = f(x)$. Then for all generizations $y' \in \mathrm{Spec}(O_y)$ of y, there exists a generization x' of x such that $f(x') = y'$.

Proof. We may assume $X = \mathrm{Spec}(O_x)$ and $Y = \mathrm{Spec}(O_y)$. Let F be the given O_X-Module. By (1.6), F is faithfully flat over O_y, so the assertion follows from (2.4)

Proposition (2.7) (Le sorite for flat morphisms). -

(i) An open immersion is flat (resp. quasi-flat).

(ii) The composition of flat (resp. faithfully flat) morphisms is flat (resp. faithfully flat).

(iii) Any base extension of a flat (resp. faithfully flat, quasi-flat, quasi-faithfully flat) morphism is flat (resp. faithfully flat, quasi-flat, quasi-faithfully flat).

(iv) The product of flat (resp. faithfully flat) morphisms is flat (resp. faithfully flat).

Proof. Assertion (i) is trivial; (ii) follows from (1.5,(iii)); (iii), from (1.5,(ii)) and (II,2.7); and (iv), from (ii) and (iii).

Proposition (2.8). - Let X and Y be locally noetherian schemes, $f : X \longrightarrow Y$ a finite morphism and F a coherent O_X-Module. If F is flat over $y \in Y$, then f_*F is locally free at y.

Proof. Since f is affine, $(f_*F)_y$ is equal to $M = \Gamma(f^{-1}(y),F)$. By (2.2), M is flat over O_y. Further, M is finite over the noetherian local ring O_y. Therefore, by (III,5.8), $(f_*F)_y$ is free.

Definition (2.9). - Let X be a scheme and Y a closed subscheme of X. The codimension of Y in X, denoted $\mathrm{codim}(Y,X)$, is defined as the infimum of the integers $\dim(O_{X,y})$ as y runs through Y.

Proposition (2.10). - Let $f : X \longrightarrow Y$ be a surjective morphism of locally noetherian schemes, Y' a closed irreducible subscheme of Y and X' an irreducible component of $f^{-1}(Y')$. Then:

(i) If $f|_{X'}: X' \longrightarrow Y'$ is generically surjective, then
$\text{codim}(X',X) \leq \text{codim}(Y',Y)$.

(ii) If f is quasi-flat, then $f|_{X'}$ is generically surjective and
$\text{codim}(X',X) = \text{codim}(f^{-1}(Y'),X) = \text{codim}(Y',Y)$.

Proof. Let z be the generic point of Y' and w the generic point of X'. By definition, $\text{codim}(Y',Y) = \dim(O_{Y,z})$; by (III,1.7),

$$\dim(O_{X,w}) \leq \dim(O_{Y,z}) + \dim(O_{X,w} \otimes_{O_{Y,z}} k(z)) \leq \dim(O_{Y,z});$$

whence (i).

Suppose f is quasi-flat. Then, by (2.6), $f(w)$ has no generization; hence $f(w) = z$. Part (ii) now results from the following proposition.

Proposition (2.11). - Let $\varphi : A \longrightarrow B$ be a local homomorphism of noetherian rings, m the maximal ideal of A and $k = A/m$. Assume that either of the following hypotheses holds:

(a) There exists a finite nonzero B-module M which is flat over A.

(b) For all primes p of A not equal to m and all minimal (essential) primes q of pB, $\varphi^{-1}(q) \neq m$.

Then $\dim(B) = \dim(A) + \dim(B \otimes_A k)$.

Proof. Assume (a) and let q be any minimal prime of pB. If $\varphi^{-1}(q) = m$, then the composition $A \longrightarrow B \longrightarrow B_q$ is a local homomorphism. By (2.3) and (1.5), M_q is flat over A; so, by (1.6), M_q is faithfully flat over A. Hence, by (2.4), there exists a prime q' of B_q such that $\varphi^{-1}(q') = p$. Thus, $qB_q \underset{\neq}{\supset} q' \supset pB$, contradicting minimality of q. Therefore (b) holds.

Assume (b). If $\dim(A) = 0$, then m is the nilradical of A by (II,4.7). Hence, mB is contained in the nilradical n of B. So, $\dim(B) = \dim(B/nB)$ and the formula holds.

Let $\dim(A) > 0$. Let $\{q_i\}$ be the set of minimal primes of B and $p_i = \varphi^{-1}(q_i)$. Suppose $p_i = m$ for some i. Since $\dim(A) > 0$, there exists a prime p of A not equal to m. Then $q_i \supset pB$ and, since q_i is a minimal prime of B, it is a fortiori a minimal prime of pB, contradicting (b). Hence $p_i \neq m$ for all i.

Let $\{p'_j\}$ be the set of minimal primes of A. Since $\dim(A) > 0$, $p'_j \neq m$. Since A and B are noetherian, they have only a finite number of minimal primes by (II,3.7). Hence, by (III,1.5), there exists an element $x \in m$, $x \notin p_i$ and $x \notin p'_j$ for all i, j. Let $A' = A/xA$, $B' = B/xB$. By (III,1.6), $\dim(B') =$ $= \dim(B)-1$ and $\dim(A') = \dim(A)-1$. Moreover, it is clear that $\dim(B\otimes_A k) = \dim(B'\otimes_{A'} k)$ and that (b) holds for $\varphi : A' \longrightarrow B'$. Hence, the formula results by induction.

3. The local criterion of flatness

Lemma (3.1). - Let $A \longrightarrow B$ be a homomorphism of rings and M an A-module. Then the following conditions are equivalent:

(i) $\mathrm{Tor}_1^A(M,N) = 0$ for all B-modules N.

(ii) $M\otimes_A B$ is a flat B-module and $\mathrm{Tor}_1^A(M,B) = 0$.

Proof. Dualized, (IV,2 2) yields the spectral sequence of a composite right-exact functor: $E^2_{pq} = L_p S(L_q T(M)) \Longrightarrow E_{p+q} = L_{p+q}(S\circ T)(M)$. With $S = -\otimes_B N$, $T = -\otimes_A B$ and $S\circ T = -\otimes_A N$, the exact sequence of terms of low degree ([2],XV,5.12a) is

$$N\otimes_B \mathrm{Tor}_1^A(M,B) \longrightarrow \mathrm{Tor}_1^A(M,N) \longrightarrow \mathrm{Tor}_1^B(M\otimes_A B,N) \longrightarrow 0,$$

and the equivalence follows easily.

Theorem (3.2). - Let A be a ring, I an ideal of A and M an A-module. Consider the following conditions:

(i) M is a flat A-module.

(ii) $M\otimes_A A/I$ is a flat (A/I)-module and $\mathrm{Tor}_1^A(M,A/I) = 0$.

(ii') $M\otimes_A A/I$ is a flat (A/I)-module and the canonical homomorphism $I\otimes_A M \longrightarrow IM$ is an isomorphism.

(iii) $\mathrm{Tor}_1^A(M,N) = 0$ for all A-modules N annihilated by I.

(iii') $\mathrm{Tor}_1^A(M,N) = 0$ for all A-modules N annihilated by I^s for some s (depending on N)

(iv) $M\otimes_A (A/I^s)$ is a flat (A/I^s)-module for all s.

(v) $M\otimes_A (A/I)$ is a flat (A/I)-module and $\gamma : \mathrm{gr}_I^0(M)\otimes_{A/I}\mathrm{gr}_I^*(A) \longrightarrow \mathrm{gr}_I^*(M)$ is an isomorphism.

Then the following implications hold:

$$(i) \Longrightarrow (ii) \Longleftrightarrow (ii') \Longleftrightarrow (iii) \Longleftrightarrow (iii') \Longrightarrow (iv) \Longleftrightarrow (v).$$

Further, suppose that I is nilpotent or that the following three conditions hold: A is noetherian; there exists a noetherian A-algebra B such that M is a finite B-module; and $IB \subset \mathrm{rad}(B)$. Then (iv) implies (i) and, hence, all the conditions are equivalent.

Proof. By (1.5), (i) implies (ii) and, by (1.8), (ii) is equivalent to (ii'). By (3.1) with $B = A/I$, (ii) is equivalent to (iii) and, by (3.1) with $B = A/I^s$, (iii') implies (iv).

The implication (iii')$\Longrightarrow$(iii) is trivial. Assume (iii). Let N be annihilated by I^s and consider the exact sequence

$$\mathrm{Tor}_1^A(M,IN) \longrightarrow \mathrm{Tor}_1^A(M,N) \longrightarrow \mathrm{Tor}_1^A(M,N/IN)$$

since IN is annihilated by I^{s-1} and N/IN is annihilated by I, the two end terms may be assumed zero by induction on s. Then $\mathrm{Tor}_1^A(M,N) = 0$ and thus (iii) is equivalent to (iii').

Consider the diagram

$$\begin{array}{ccccccccc}
 & & I^{s+1}\otimes_A M & \longrightarrow & I^s\otimes_A M & \longrightarrow & gr_I^s(A)\otimes_{A/I} gr_I^0(M) & \longrightarrow & 0 \\
 & & \downarrow{\scriptstyle \theta_{s+1}} & & \downarrow{\scriptstyle \theta_s} & & \downarrow{\scriptstyle \gamma_s} & & \\
0 & \longrightarrow & I^{s+1}M & \longrightarrow & I^sM & \longrightarrow & gr_I^s(M) & \longrightarrow & 0
\end{array}$$

Assume (iii'). Then, by (1.8), θ_s and θ_{s+1} are isomorphisms. Thus, for all $s > 0$, γ_s is an isomorphism; hence, $\gamma = \oplus\,\gamma_s$ is an isomorphism. Furthermore, by (iii') $\Longrightarrow$ (ii), $M\otimes_A(A/I)$ is a flat (A/I)-module. Thus, (iii') implies (v).

If θ_{s+1} is an isomorphism, the map $I^{s+1}\otimes_A M \longrightarrow I^s\otimes_A M$ is injective. If further (v) holds, γ_s is an isomorphism; so by the five lemma, θ_s is an isomorphism. If I is nilpotent, then θ_{s+1} is an isomorphism for large s; hence, if (v) also holds, descending induction yields (ii').

Fix $n > 0$ and replace A by A/I^n, I by I/I^n and M by M/I^nM to obtain conditions $(i)_n$, $(ii)_n$, $(iii)_n$, $(iv)_n$ and $(v)_n$. The implication (iv) $\Longrightarrow (i)_n$ is trivial; $(i)_n \Longrightarrow (v)_n$, proved. Observe

$$gr^s_{(I/I^n)}(M/I^nM) = \begin{cases} gr_I^s(M) & \text{for } s < n \\ 0 & \text{for } s \geqslant n \end{cases} ;$$

hence, if $(v)_n$ holds for all n, then (v) holds. Therefore, (iv) implies (v).

Since I/I^n is nilpotent, $(v)_n$ implies $(ii')_n$. However, the implications (v) $\Longrightarrow (v)_n$ and $(ii')_n \Longrightarrow (iv)_n$ are proved, and, clearly, if $(iv)_n$ holds for all n, then (iv) holds. Hence, (v) implies (iv).

It remains to prove the implication (iv) $\Longrightarrow$ (i) under the following conditions: A is noetherian; there exists a noetherian A-algebra B such that M is a finite B-module; and $IB \subset \mathrm{rad}(B)$.

Let $N' \longrightarrow N$ be an injection of finite A-modules and consider the injection $h : N'/(I^rN \cap N') \longrightarrow N/I^rN$. Then $h \otimes \mathrm{id}_M$ may be written in the form

$h \otimes \mathrm{id}_{M \otimes (A/I^r)} : (N'/(I^rN \cap N') \otimes_{(A/I^r)} (M \otimes_A (A/I^r)) \to (N/I^rN) \otimes_{(A/I^r)} (M \otimes_A A/I^r)$, thus $h \otimes \mathrm{id}_M$ is injective by (iv). By the Artin-Rees lemma (II,1.14), there exists an integer $k \geq 0$ such that $I^{r-k}(N' \cap I^kN) = N' \cap I^rN$ for all $r > 0$. Let M' be the image of $(N' \cap I^kN) \otimes_A M$ in $N' \otimes_A M$. Then $h \otimes \mathrm{id}_M$ becomes $g : N' \otimes_A M/I^{r-k}M' \longrightarrow N \otimes_A M/I^r(N \otimes_A M)$. The filtrations $(I^{r-k}M')$ and $(I^r(N' \otimes_A M))$ induce the same topology on $N' \otimes_A M$; hence, by (II,1.9) and 1.8), $\hat{g} : (N' \otimes_A M)^\wedge \longrightarrow (N \otimes_A M)^\wedge$ is injective. Therefore, by Krull's intersection theorem (II,1.15), $N' \otimes_A M \longrightarrow N \otimes_A M$ is injective. Hence, it follows from (1.7) and (1.8) that M is flat, completing the proof of the local criterion.

<u>Proposition (3.3)</u>. - Let $A \longrightarrow B$ be a homomorphism of noetherian rings, I an ideal of A and I' an ideal of B such that $IB \subset I' \subset \mathrm{rad}(B)$. Let M be a finite B-module and $\hat{M} = \varprojlim M/I'^nM$. Then the following conditions are equivalent:

(i) M is flat over A.

(ii) $\hat{M}$ is flat over A.

(iii) $\hat{M}$ is flat over $\hat{A}$.

<u>Proof</u>. Since $\hat{B}$ is faithfully flat over B (1.10), the functor $- \otimes_A M$ is exact if and only if $- \otimes_A M \otimes_B \hat{B}$ is exact. However, by (II,1.18), $- \otimes_A \hat{M} = - \otimes_A M \otimes_B \hat{B}$. Hence (i) and (ii) are equivalent.

By (II,1.18), $\hat{M}$ is a finite $\hat{B}$-module; by (II,1.22), $\hat{B}$ is a noetherian A-(resp. $\hat{A}$-) algebra, and A and $\hat{A}$ are both noetherian rings; and, by (II,1.23), $I\hat{B} \subset \mathrm{rad}(\hat{B})$. Since $A/I^n \cong \hat{A}/\hat{I}^n$ by (II,1.19), the equivalence of (i) and (iv) of the local criterion (3.2), yields the equivalence of (ii) and (iii).

Proposition (3.4). - Let $R \longrightarrow A$ and $A \longrightarrow B$ be local homomorphisms of noetherian rings and let M be a finite B-module. Suppose A is flat over R. Then M is flat over A if (and only if) the following two conditions hold:

(a) M is flat over R.

(b) $M \otimes_R k$ is flat over $A \otimes_R k$ where $k = R/m$ and m is the maximal ideal.

Proof. The implication (i) $\Longrightarrow$ (v) of the local criterion applied to M yields $(M/IM) \otimes_k gr_m^*(R) \xrightarrow{\sim} gr_I^*(M)$ where $I = mA$, and to A yields $(A/I) \otimes_k gr_m^*(R) \xrightarrow{\sim} gr_I^*(A)$. Therefore, by (v) $\Longrightarrow$ (i) of the local criterion, M is flat over A.

Proposition (3.5). - Let $A \longrightarrow B$ be a local homomorphism of noetherian rings. Let M be a finite B-module, m the maximal ideal of A and $k = A/m$. Assume the following conditions hold:

(a) A is a regular local ring.

(b) M is a Cohen-Macaulay B-module.

(c) $\dim_B(M) = \dim(A) + \dim_{B \otimes_A k}(M \otimes_A k)$.

Then M is flat over A.

Proof. Since k is a field, $M \otimes_A k$ is flat over k. So, by (ii) $\Longrightarrow$ (i) of the local criterion, it suffices to prove $\mathrm{Tor}_1^A(M,k) = 0$. Let $x_1, \ldots, x_r$ be regular parameters of A where $r = \dim(A)$. Then, by (c),

$$\dim_B(M/(x_1M + \ldots + x_rM)) = \dim_B(M) - \dim(A).$$

Hence, by the Cohen-Macaulay theorem (III,4.3), $(x_1, \ldots, x_r)$ is an M-regular sequence.

Let $M_i = M/(x_1M + \dots + x_iM)$ and $A_i = A/(x_1A + \dots + x_iA)$. We prove $\mathrm{Tor}_1^A(M,A_i) = 0$ by induction on i. If $i = 0$, then $A_0 = A$ is A-flat. If $i \geq 0$, then the exact sequence (III,4.11) $0 \longrightarrow A_i \xrightarrow{x_{i+1}} A_i \longrightarrow A_{i+1} \longrightarrow 0$ yields an exact sequence

$$\mathrm{Tor}_1^A(M,A_i) \longrightarrow \mathrm{Tor}_1^A(M,A_{i+1}) \longrightarrow M_i \xrightarrow{x_{i+1}} M_i .$$

By induction, $\mathrm{Tor}_1^A(M,A_i) = 0$ and, by M-regularity, multiplication by x_{i+1} is injective; hence, $\mathrm{Tor}_1^A(M,A_{i+1}) = 0$.

<u>Corollary (3.6)</u>. - Let $A \longrightarrow B$ be a quasi-finite, (cf VI,2.1), local homomorphism of regular local rings having the same dimension. Then B is flat over A.

<u>Proof</u>. Let k be the residue field of A. Since B is quasi-finite over A, $\dim(B \otimes_A k) = 0$ (II,4.5 and 4.7). By (III,4.12), B is Cohen-Macaulay. Hence, (3.5) yields the assertion.

4. Constructible sets

<u>Definition (4.1)</u>. - Let X be a noetherian topological space (i.e., the closed sets satisfy the minimum condition). A subset Z is said to be constructible if it is a finite union of locally closed subsets of X.

<u>Remark (4.2)</u>. -

(i) Open sets and closed sets are constructible.

(ii) If Z and Z' are constructible, then $Z \cup Z'$ and $Z \cap Z'$ are constructible.

(iii) If $f : Y \longrightarrow X$ is continuous and Z is constructible in X, then $f^{-1}(Z)$ is constructible in Y.

(iv) If Z is constructible in Y and Y is constructible in X, then Z is constructible in X.

<u>Lemma (4.3)</u>. - Let X be a noetherian space. A subset Z is constructible if and only if the following condition holds: For all closed irreducible subsets Y such that $Z \cap Y$ is dense in Y, there exists a nonempty set V in $Z \cap Y$ which is open in Y.

<u>Proof</u>. Suppose Z is constructible; say, $Z = \bigcup_{i=1}^{n} (V_i \cap F_i)$ with the V_i open and the F_i closed. Let Y be a closed irreducible subset such that $Z \cap Y$ is dense in Y. Then $Z \cap Y = \cup (V_i' \cap F_i')$ where $V_i' = V_i \cap Y$ and $F_i' = F_i \cap Y$. Now, the dense subset $Z \cap Y$ of Y is contained in the closed subset $\cup F_i'$; so, $Y = \cup F_i'$. However, Y is irreducible; so, for some j, $F_j' = Y$ and $V_j' = V_j' \cap F_j' \subset Z \cap Y$.

Conversely, suppose the condition is satisfied. Let S be the family of closed subsets Y of X such that $Z \cap Y$ is not constructible. Suppose S is nonempty and let X' be a minimal element of S. Replacing X' by X, we may assume $Z \cap Y$ is constructible for all proper closed subsets Y.

Suppose $X = X_1 \cup X_2$ where $X_1 ; X_2$ are proper closed subsets. Then each $Z \cap X_i$ is constructible; hence, $Z = (Z \cap X_1) \cup (Z \cap X_2)$ is constructible.

Suppose X is irreducible. If the closure $\bar{Z}$ of Z is a proper subset, then $Z = Z \cap \bar{Z}$ is constructible. If $\bar{Z} = X$, then, by hypothesis, there exists a nonempty open set V in Z. Then $F = X - V$ is a proper closed subset; so, $Z = V \cup (F \cap Z)$ is constructible.

<u>Lemma (4.4)</u>. - Let X be a noetherian space such that every closed irreducible subset has a generic point. Let Z be a constructible subset of X and $x \in Z$. Then Z is a neighborhood of x if (and only if) every generization x' of x is in Z.

Proof. By noetherian induction, we may assume that, for every proper closed subset Y of X which contains x, $Y \cap Z$ is a neighborhood of x in Y. Suppose $X = X_1 \cup X_2$ where X_1 and X_2 are proper closed subsets. For $i = 1,2$, if $x \in X_i$, then, by assumption, there exists an open set V_i of X_i such that $x \in V_i \subset X_i \cap Z$; if $x \notin X_i$, set $V_i = \emptyset$. Let $F_i = X_i - V_i$, $F = F_1 \cup F_2$ and $V = X-F$. Then V is a neighborhood of x and $V \subset V_1 \cup V_2 \subset Z$; so, Z is a neighborhood of x.

Suppose X is irreducible. If x' is its generic point, then, by hypothesis $x' \in Z$; whence, $\bar{Z} = X$. So, by (4.3), there exists a subset V of Z which is open. If $x \in V$, the proof is complete. If $x \notin V$, let $Y = X-V$. Then, Y is a proper closed subset of X and $x \in Y$. Hence, by assumption, $Y \cap Z$ is a neighborhood of x in Y. Let F be the closure of $X-Z$ in X. Then F is also the closure of $X-Z$ in $X-V = Z$; so, $x \notin F$. Let $V' = X-F$. Then V' is a neighborhood of x contained in Z and thus Z is a neighborhood of x.

Proposition (4.5). - Let X be a locally noetherian space such that every closed irreducible subset has a generic point. Then a subset V of X is open if (and only if) the following two conditions are satisfied for all $x \in V$:

(a) V contains every generization of x.

(b) $V \cap \overline{\{x\}}$ is a neighborhood of x in $\overline{\{x\}}$.

Proof. The assertion being local, we may assume X is noetherian. Then, by (4.3), V is constructible; hence, by (4.4), V is open.

Theorem (4.6) (Chevalley). - Let $f : X \longrightarrow Y$ be a morphism of finite type of noetherian schemes. Let Z be a constructible subset of X. Then $f(Z)$ is constructible.

Proof. Let $Z = \bigcup_{i=1}^{n} Z_i$ where the Z_i are locally closed. Give each Z_i the (unique) induced, reduced subscheme structure. Since X is a noetherian space, the immersions $Z_i \hookrightarrow X$ are of finite type. Replacing X by $\amalg Z_i$, we may therefore assume $Z = X$ and X is reduced.

Let T be a closed irreducible subset of Y such that $T \cap f(X)$ is dense in T; in view of (4.3), it suffices to prove that $T \cap f(X)$ contains an open set of T. Since $T \cap f(X) = f(f^{-1}(T))$, if we replace Y by T and X by $f^{-1}(T)$, given their reduced subscheme structures, we may assume that $f(X)$ is dense in Y and that Y is reduced and irreducible.

We clearly may assume Y is affine. Then $X = \cup X_i$ with X_i affine and irreducible. Since Y is irreducible, $f(X_j)$ is dense in Y for some j. Hence, replacing X by X_j, we may assume X is affine, reduced and irreducible.

Let $Y = \mathrm{Spec}(A)$ and $X = \mathrm{Spec}(B)$ where A and B are integral domains and B is of finite type over A. Since $f(X)$ is dense in Y, we may assume A is contained in B. It now remains to show that there exists a nonzero element $g \in A$ such that, for all primes p of A such that $g \notin p$, there exists a prime P of B such that $p = A \cap P$. Take $g \in A$ and $C = A[T_1,\ldots,T_n]$ as provided by the lemma below. Then pC_g is prime in C_g; so, since B_g is integral over C_g, there exists a prime P' of B_g lying over pC_g by (III,2.2). Let $P = P' \cap B$; then $P \cap A = p$.

Lemma (4.7). - Let A be a domain and B an A-algebra of finite type which contains A. Then there exists a nonzero element g of A and a subalgebra C of B isomorphic to a polynomial algebra $A[t_1,\ldots,t_m]$ such that B_g is integral over C_g.

Proof. Let $S = A-\{0\}$ and $K = S^{-1}A$. Then, by (III,2.5), there exist elements $T_1,\dots,T_n \in S^{-1}B$, algebraically independent over K, such that $S^{-1}B$ is integral over the polynomial algebra $K[T_1,\dots,T_n]$. There exists $g \in S$ such that $T_i = t_i/g$ with $t_i \in B$ and such that the integral equations of generators $z_1,\dots,z_n$ of $S^{-1}B$ over K have coefficients of the form c/g with $c \in A$. Then B_g is integral over $A[t]_g$.

Proposition (4.8). - Let X and Y be locally noetherian schemes and $f : X \longrightarrow Y$ a morphism locally of finite type. Let x be a point of X and $y = f(x)$. If V is a neighborhood of x, then $f(V)$ is a neighborhood of y if (and only if), for all generizations y' of y, there exists a generization x' of x such that $f(x') = y'$.

Proof. We may assume that X, Y are affine and noetherian and that V is open. By (4.6), $f(V)$ is constructible; so, by (4.4), $f(V)$ is a neighborhood of y.

5. Flat morphisms and open sets

Theorem (5.1). - Let X and Y be locally noetherian schemes and $f : X \longrightarrow Y$ a morphism locally of finite type. If f is quasi-flat, then f is open.

Proof. Let U be an open set of X and $y = f(x)$ a point of $f(U)$. By (2.6), for any generization y' of y, there exists a generization x' of x such that $f(x') = y'$; hence, by (4.8), f is open.

Theorem (5.2) (Lemma of generic flatness). - Let A be a noetherian domain, B an A-algebra of finite type and M a finite

B-module. Then there exists a nonzero element f of A such that M_f is free over A_f.

<u>Proof</u>. If K is the quotient field of A, then $B\otimes_A K$ is a K-algebra of finite type and $M\otimes_A K$ is a $(B\otimes_A K)$-module of finite type. Let $n = \dim(M\otimes_A K)$.

If $n < 0$, then $M\otimes_A K = 0$. Let $\{g_1,\ldots,g_n\}$ be a set of generators of M over B. There exists a nonzero element f of A such that $fg_i = 0$ for all i. Then $M_f = 0$.

By (II,3.7), there exists a filtration of B-modules

$$M = M_0 \supset \ldots. \supset M_q = 0$$

such that $M_i/M_{i+1} \cong B/p_i$ for suitable primes p_i of B. Suppose there exist elements $f_i \in A$ such that the $(M_i/M_{i+1})_{f_i}$ are free over A_{f_i}. If $f = \Pi f_i$, then M_f is free over A_f. Hence, we may assume M is of the form B/p. Further, replacing B by B/p, we may assume B is a domain. Let I be the annihilator of the A-module B. If $0 \neq g \in I$, then $B_g = 0$; so, $B\otimes_A K = 0$.

Assume $n = \dim(B\otimes_A K)$ is not zero. Then, by the above paragraph, $A \longrightarrow B$ is injective. By (4.7), there exists a nonzero element g of A and a polynomial algebra $C = A[T_1,\ldots,T_r]$ contained in B such that B_g is integral over C_g. Replacing A by A_g and B by B_g, we may assume B is integral over C. Hence, by (III,2.2), $n = \dim(C\otimes_A K)$. There exists an exact sequence of C-modules of the form

$$0 \longrightarrow C^m \longrightarrow B \longrightarrow N \longrightarrow 0$$

where $m = \dim_{K(T)}(B\otimes_A K(T))$. It follows that $\dim(N\otimes_A K) < n$. Hence, by induction, there exists a nonzero element h of A such that N_h is a free A_h-module. Therefore, B_h is a free A_h-module and the proof of (5.2) is complete.

<u>Lemma (5.3)</u>. - Let A be a noetherian ring, B an A-algebra of finite type and M a finite B-module. Let p be a prime of B and q the trace of p in A. Suppose M_p is flat over A_q (or, equivalently, over A). Then there exists a nonzero element g of A such that:

(i) $(M/qM)_g$ is flat over A/q.

(ii) $\mathrm{Tor}_1^A(M,A/q)_g = 0$

<u>Proof</u>. The lemma (5.2) of generic flatness, applied to A/q, yields an $f \in A-q$ such that $(M/qM)_f$ is flat over A/q. By hypothesis $0 = \mathrm{Tor}_1^A(M_p,A/q) = \mathrm{Tor}_1^A(M,A/q)_p$. Since $\mathrm{Tor}_1^A(M,A/q)$ is a finite B-module, there exists an element h of $B-p$ such that $\mathrm{Tor}_1^A(M,A/q)_h = 0$. Then (i) and (ii) hold for $g = fh$.

<u>Lemma (5.4)</u>. - Under the assumptions of (5.3), if p' is a prime of B containing p such that $g \notin p'$, then $M_{p'}$ is flat over A_q (or, equivalently, over A).

<u>Proof</u>. By (5.3, (i)) and (2.3), $M_{p'}/qM_{p'}$ is flat over A/q and, by (5.3,(ii)), $0 = \mathrm{Tor}_1^A(M,A/q)_{p'} = \mathrm{Tor}_1^A(M_{p'},A/q)$. Hence, the local criterion (3.2), applied to the A-algebra $B_{p'}$, the $B_{p'}$-module $M_{p'}$ and the ideal q, yields the assertion.

<u>Theorem (5.5)</u>. - Let X and Y be locally noetherian schemes and $f : X \longrightarrow Y$ a morphism locally of finite type. Let F be a coherent O_X-Module and U the set of points $x \in X$ such that F_x is flat over $O_{f(x)}$. Then U is open.

<u>Proof</u>. Since generization corresponds to localization, it follows from (2.3), (5.3) and (5.4) that the two conditions of (4.5) hold; hence, U is open.

Chapter VI - Étale Morphisms

1. Differentials

<u>Definition (1.1)</u>. - Let k be a ring, A a k-algebra and M an A-module. The module of <u>k-derivations of</u> A <u>in</u> M, denoted $Der_k(A,M)$, is defined as the set of all maps $D : A \longrightarrow M$ satisfying the following two conditions:

(a) D is k-linear.

(b) $D(fg) = fD(g) + gD(f)$ for all $f, g \in A$.

<u>Remark (1.2)</u>. - Let k be a ring, A a k-algebra, M an A-module and $D : A \longrightarrow M$ a Z-linear map. Then:

(i) If D satisfies (b), then D satisfies (a) if and only if $D(f) = 0$ for all $f \in k$.

(ii) $Der_k(A,M)$ is a functor in M.

<u>Definition (1.3)</u>. - Let k be a ring and A a k-algebra. Suppose that the functor $M \longmapsto Der_k(A,M)$ is represented by the pair $(d_{A/k}, \Omega^1_{A/k})$; namely, suppose that $\Omega^1_{A/k}$ is an A-module, that $d_{A/k} \in Der_k(A, \Omega^1_{A/k})$ and that, given any A-module M and any k-derivation $D : A \longrightarrow M$, there exists a unique A-homomorphism $w : \Omega^1_{A/k} \longrightarrow M$ such that the following diagram commutes:

$$\begin{array}{ccc} A & \xrightarrow{d_{A/k}} & \Omega^1_{A/k} \\ {\scriptstyle D}\downarrow & \swarrow{\scriptstyle w} & \\ M & & \end{array}$$

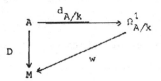

(or, equivalently, that the map of functors $Hom_A(\Omega^1_{A/k}, -) \longrightarrow Der_k(A,-)$, induced by $d_{A/k}$, is an isomorphism). By "abstract nonsense", the pair $(d_{A/k},\Omega^1_{A/k})$ is easily seen to be unique up to unique isomorphism. The A-module $\Omega^1_{A/k}$ is called the module of <u>1-differentials of</u> A <u>over</u> k; $d_{A/k}$, the <u>exterior differential of</u> A <u>over</u> k; and $(d_{A/k},\Omega^1_{A/k})$, the <u>differential pair of</u> A <u>over</u> k.

<u>Proposition (1.4)</u>. - Let k be a ring and $A = k[T_\alpha]$ a polynomial algebra (in possibly infinitely many variables). Let Ω be the free A-module on the symbols dT_α and $d : A \longrightarrow \Omega$ the derivation defined by $dP(T) = \sum \frac{\partial P}{\partial T_\alpha} dT_\alpha$. Then (d,Ω) is the differential pair of A over k.

<u>Proof</u>. Let M be an A-module, $D \in \mathrm{Der}_k(A,M)$ and define $w : \Omega \longrightarrow M$ by $w(dT_\alpha) = D(T_\alpha)$. Then $w(dP(T)) = \sum \frac{\partial P}{\partial T_\alpha} w(dT_\alpha) = D(P(T))$; whence, the assertion.

<u>Remark (1.5)</u>. - Let
$$\begin{array}{ccc} A & \xrightarrow{\varphi} & B \\ \uparrow & & \uparrow i \\ k & \longrightarrow & k' \end{array}$$
be a commutative diagram of commutative rings and suppose the differential pairs $(d_{A/k},\Omega^1_{A/k})$, $(d_{B/k},\Omega^1_{B/k})$ and $(d_{B/k'},\Omega^1_{B/k'})$ exist. Then, since $d_{B/k'} \in \mathrm{Der}_k(B,\Omega^1_{B/k})$, there exists a unique B-homomorphism $v_{B/k'/k}: \Omega^1_{B/k} \longrightarrow \Omega^1_{B/k'}$ such that $d_{B/k'} = v_{B/k'/k} \circ d_{B/k}$. Furthermore, since $d_{B/k} \circ \varphi \in \mathrm{Der}_k(A,\Omega^1_{B/k})$, there exists a unique A-homomorphism $w : \Omega^1_{A/k} \longrightarrow \Omega^1_{B/k}$ such that $w \circ d_{A/k} = d_{B/k} \circ \varphi$; whence a B-homomorphism $u_{B/A/k}: \Omega^1_{A/k} \otimes_A B \longrightarrow \Omega^1_{B/k}$ such that the following diagram commutes:

$$\begin{array}{ccccc} \Omega^1_{A/k} \otimes_A B & \xrightarrow{u_{B/A/k}} & \Omega^1_{B/k} & \xrightarrow{v_{B/k'/k}} & \Omega^1_{B/k'} \\ {\scriptstyle d_{A/k} \otimes_k i}\uparrow & & \uparrow{\scriptstyle d_{B/k}} & & \uparrow{\scriptstyle d_{B/k'}} \\ A \otimes_k k' & \xrightarrow{\varphi \otimes i} & B & \xrightarrow{id_B} & B \end{array}.$$

<u>Theorem (1.6)</u>. - Let k be a ring, $\varphi : A \longrightarrow B$ a k-algebra homomorphism. If the differential pairs exist, then there exists a canonical exact sequence of B-modules

$$\Omega^1_{A/k} \otimes_A B \xrightarrow{u_{B/A/k}} \Omega^1_{B/k} \xrightarrow{v_{B/A/k}} \Omega^1_{B/A} \longrightarrow 0$$

<u>Proof</u>. If M is a B-module, then the sequence

$$0 \longrightarrow \mathrm{Der}_A(B,M) \longrightarrow \mathrm{Der}_k(B,M) \longrightarrow \mathrm{Der}_k(A,M)$$

is easily seen exact in view of (1.2,(i)). It follows that the sequence

$$0 \longrightarrow \mathrm{Hom}_B(\Omega^1_{B/A},M) \longrightarrow \mathrm{Hom}_B(\Omega^1_{B/k},M) \longrightarrow \mathrm{Hom}_B(\Omega^1_{A/k} \otimes_A B,M)$$

is exact. Therefore, the following lemma completes the proof.

<u>Lemma (1.7)</u>. - Let B be a ring. A sequence $N' \xrightarrow{f} N \xrightarrow{g} N'' \longrightarrow 0$ of B-modules is exact if (and only if) the sequence $0 \longrightarrow \mathrm{Hom}(N'',M) \longrightarrow \mathrm{Hom}(N,M) \longrightarrow \mathrm{Hom}(N',M)$ is exact for all B-modules M.

<u>Proof</u>. Since $0 \longrightarrow \mathrm{Hom}(N'',\mathrm{coker}(g)) \longrightarrow \mathrm{Hom}(N,\mathrm{coker}(g))$ is exact, the canonical map $N'' \longrightarrow \mathrm{coker}(g)$ is 0; so, g is surjective. Since $\mathrm{Hom}(N'',N'') \longrightarrow \mathrm{Hom}(N,N'') \longrightarrow \mathrm{Hom}(N',N'')$ is exact, $\mathrm{id}_N \circ g \circ f = 0$. So there exists a canonical map $h : \mathrm{coker}(f) \longrightarrow N''$. Since $\mathrm{Hom}(N'',\mathrm{coker}(f)) \longrightarrow \mathrm{Hom}(N,\mathrm{coker}(f)) \longrightarrow \mathrm{Hom}(N',\mathrm{coker}(f))$ is exact, the canonical map $N'' \longrightarrow \mathrm{coker}(f)$ yields an inverse to h, completing the proof.

<u>Theorem (1.8)</u>. - Let k be a ring, A a k-algebra, I an ideal of A and B = A/I. Suppose the differential pair of A over k exists. Then the differential pair of B over k exists and there

exists a canonical exact sequence of B-modules

$$I/I^2 \xrightarrow{\delta} \Omega^1_{A/k} \otimes_A B \longrightarrow \Omega^1_{B/k} \longrightarrow 0$$

where δ is induced by $d_{A/k}$.

__Proof__. Let M be a B-module. Then the sequence

$$0 \longrightarrow \mathrm{Der}_k(B,M) \longrightarrow \mathrm{Der}_k(A,M) \longrightarrow \mathrm{Hom}_B(I/I^2,M)$$

is easily seen exact. However, the sequence
$0 \longrightarrow \mathrm{Hom}_B(\mathrm{coker}(\delta),M) \longrightarrow \mathrm{Der}_k(A,M) \longrightarrow \mathrm{Hom}_B(I/I^2,M)$ is also exact.
Therefore, $\Omega^1_{B/k}$ exists and is equal to coker(δ).

__Theorem (1.9)__. - Let k be a ring and B a k-algebra. Then the differential pair $(d_{B/k},\Omega^1_{B/k})$ exists.

__Proof__. Since B is a quotient of some polynomial algebra $A = k[T]$, the assertion follows from (1.4) and (1.8).

__Lemma (1.10)__. - Let k be a ring, A a k-algebra, Ω an A-module and $d : A \longrightarrow \Omega$ a k-derivation. Suppose that $d(A)$ generates Ω and that there exists a map $w : \Omega \longrightarrow \Omega^1_{A/k}$ such that $d_{A/k} = w \circ d$. Then w induces an isomorphism,

$$(d,\Omega) \xrightarrow{\sim} (d_{A/k},\Omega^1_{A/k}).$$

__Proof__. Since d is a k-derivation, there exists a map $w' : \Omega^1_{A/k} \longrightarrow \Omega$ such that $d = w' \circ d_{A/k}$. Since $d(A)$ generates Ω, w' is surjective. By uniqueness, $w \circ w' = \mathrm{id}$; hence, w' is also injective.

__Proposition (1.11)__. - Let k be a ring and A a k-algebra. Then $\Omega^1_{A/k}$ is generated by the differentials $d_{A/k}(f)$ as f runs through any set of algebra generators of A over k.

Proof. Let Ω be the submodule of $\Omega^1_{A/k}$ generated by the $d_{A/k}(f)$. Then (1.10) implies that the inclusion $w : \Omega \longrightarrow \Omega^1_{A/k}$ is an isomorphism.

Proposition (1.12). - Let k be a ring, B_1, B_2 two k-algebras and $A = B_1 \otimes_k B_2$. If $d = (d_{B_1/k} \otimes id_A) + (d_{B_2/k} \otimes id_A)$ and $\Omega = (\Omega^1_{B_1/k} \otimes_{B_1} A) \oplus (\Omega^1_{B_2/k} \otimes_{B_2} A)$, then (d, Ω) is the differential pair of A over k.

Proof. By (1.11), the image of d generates Ω. By (1.5), the canonical injections $B_i \longrightarrow A$ induce maps $u_i = u_{A/B_i/k}$ and, if $w = u_1 + u_2 : \Omega \longrightarrow \Omega^1_{A/k}$, then clearly $w \circ d = d_{A/k}$. Hence, the assertion follows from (1.10).

Proposition (1.13). - Let k be a ring, B a k-algebra and $A = B \otimes_k B$. Let $p : B \otimes_k B \longrightarrow B$ be the map defined by $p(f \otimes g) = fg$, $I = \ker(p)$ and $d : B \longrightarrow I/I^2$ the k-homomorphism defined by $d(f) = 1 \otimes f - f \otimes 1$. Then d is a k-derivation, the sequence

$$0 \longrightarrow I/I^2 \xrightarrow{\delta} \Omega^1_{A/k} \otimes_A B \longrightarrow \Omega^1_{B/k} \longrightarrow 0$$

is exact and split, and $(d, I/I^2)$ is the differential pair of B over k.

Lemma (1.14). - Under the conditions of (1.13), I is generated over B (via j_1) by the elements of the form $1 \otimes f - f \otimes 1$.

Proof. Clearly, $1 \otimes f - f \otimes 1 \in I$ for all $f \in B$. If $\Sigma f_i \otimes g_i \in I$, then $\Sigma f_i g_i = 0$; so, $\Sigma f_i \otimes g_i = \Sigma (f_i \otimes 1)(1 \otimes g_i - g_i \otimes 1)$.

In (1.13), d is a derivation: $d(fg) = 1 \otimes fg - fg \otimes 1 = (1 \otimes f)(1 \otimes g - g \otimes 1) + (g \otimes 1)(1 \otimes f - f \otimes 1) = fdg + gdf$. By (1.14), $d(B)$ generates I/I^2. In view of (1.12), let $pr_2 : \Omega^1_{A/k} \otimes_A B \longrightarrow \Omega^1_{B/k}$ be the projection on the second factor and $w = pr_2 \circ \delta$. Then, since

$\circ(1\otimes f - f\otimes 1) = -df \oplus df$, it follows that $w \circ d = d_{B/k}$. Hence, (1.10) yields the assertion.

Remark (1.15). - (1.13) suggests an alternate existence proof: direct establishment of universality of $(d, I/I^2)$. Let $D : B \longrightarrow M$ be a k-derivation and define a k-homomorphism $D' : B\otimes B \longrightarrow M$ by $D'(f\otimes g) = fDg$. Then $D'((1\otimes f - f\otimes 1)(1\otimes g - g\otimes 1)) = D(fg) - fDg - gDf + 0 = 0$; hence, by (1.14), $D'(I^2) = 0$. Thus, D' induces a B-homomorphism $w : I/I^2 \longrightarrow M$ and $w(df) = w(1\otimes f - f\otimes 1) = Df$.

Example (1.16). - Let k be a ring and $B = k[T_\alpha]$ a polynomial algebra. Then $A = B\otimes_k B = k[T_\alpha, U_\beta]$. Let $U_\alpha = T_\alpha + h_\alpha$; by (1.14), I/I^2 is the B-module generated by the h_α and, by (1.13), $\delta : I/I^2 \longrightarrow \Omega^1_{B/k}$ is an isomorphism defined by $\delta(h_\alpha) = dT_\alpha$. If $P(T) \in B$, then $P(T+h) - P(T) = \sum \frac{\partial P}{\partial T_\alpha} h_\alpha + O(h^2)$ where $O(h^2) \in I^2$. Hence, as in (1.4), $\Omega^1_{B/k}$ is the free B-module generated by symbols dT_α and $dP(T) = \sum \frac{\partial P}{\partial T_\alpha} dT_\alpha$.

Proposition (1.17). - Let k be a ring, B_1, B_2 two k-algebras and $A = B_1 \times B_2$. Then the differential pair of A over k is $(d_{B_1/k} + d_{B_2/k}, \Omega^1_{B_1/k} \oplus \Omega^1_{B_2/k})$.

Proof. The assertion results formally from the fact that the category of A-modules is the direct product of the categories of B_1-modules and B_2-modules.

Proposition (1.18). - Let k be a ring, A, k' two k-algebras and $A' = A\otimes_k k'$. Then $(d_{A/k} \otimes \mathrm{id}_{A'}, \Omega^1_{A/k} \otimes_A A')$ is the differential pair of A' over k'.

Proof. By (1.11), $d_{A/k} \otimes_A \mathrm{id}_{A'} = d_{A/k} \otimes_k \mathrm{id}_{k'} : A' \longrightarrow \Omega^1_{A/k} \otimes_A A' = \Omega^1_{A/k} \otimes_k k'$ is a k'-derivation whose image generates. Furthermore, by (1.5), $d_{A'/k'} = (v_{A'/k'/k} \circ u_{A'/A/k}) \circ (d_{A/k} \otimes id_{A'})$. Hence, (1.10) yields the assertion.

Corollary (1.19). - Let k be a ring, B_1, B_2 two k-algebras and $A = B_1 \otimes_k B_2$. Then the homomorphism $j_1 : B_1 \longrightarrow A$, given $j_1(b) = b \otimes 1$, defines a canonical sequence

$$0 \longrightarrow \Omega_{B_1/k} \otimes_{B_1} A \longrightarrow \Omega^1_{A/k} \longrightarrow \Omega^1_{A/B_1} \longrightarrow 0$$

which is exact and split.

Proof. By (1.18), $\Omega^1_{A/B_1} = \Omega^1_{B_2/k} \otimes_{B_2} A$, so the assertion results immediately from (1.12).

Proposition (1.20). - Let k be a ring, A a k-algebra and σ (resp. S) a multiplicative set in k (resp. A) such that σ maps into S. Then the differential pair of $S^{-1}A$ over $\sigma^{-1}k$ is $(d, S^{-1}\Omega^1_{A/k})$ where $d(\frac{a}{s}) = (sd_{A/k}(a) - ad_{A/k}(s))/s^2$.

Proof. The image of the k-derivation $d : S^{-1}A \longrightarrow S^{-1}\Omega^1_{A/k}$ generates $S^{-1}\Omega^1_{A/k}$ by (1.11) The composition of the natural homomorphism $h : A \longrightarrow S^{-1}A$ with $d_{S^{-1}A/\sigma^{-1}k}$ is a k-derivation; so there exists an A-homomorphism $w : \Omega^1_{A/k} \longrightarrow \Omega^1_{S^{-1}A/\sigma^{-1}k}$ such that $d_{S^{-1}A/\sigma^{-1}k} \circ h = w \circ d_{A/k}$. Since $\Omega^1_{S^{-1}A/\sigma^{-1}k}$ is an $S^{-1}A$-module, w may be extended to $w : S^{-1}\Omega^1_{A/k} \longrightarrow \Omega^1_{S^{-1}A/\sigma^{-1}k}$ such that $w \circ d = d_{S^{-1}A/\sigma^{-1}k}$. Hence, the assertion results from (1.10).

Remark (1.21). - In geometric terms, this discussion may be reinterpreted as follows. Let X be an S-scheme. By (1.20) and (1.9), there exists a canonical pair $(d_{X/S}, \Omega^1_{X/S})$ consisting of a quasi-coherent O_X-Module $\Omega^1_{X/S}$ and a map $d_{X/S} : O_X \longrightarrow \Omega^1_{X/S}$ defined as follows: for each open affine subset $V = \mathrm{Spec}(k)$ of S and for each open affine subset $U = \mathrm{Spec}(A)$ of X lying over V, $\Omega^1_{X/S}|U = (\Omega^1_{A/k})^\sim$ and $d_{X/S}|U = (d_{A/k})^\sim$. The O_X-Module $\Omega^1_{X/S}$ is called the sheaf of 1-differential forms and the map $d_{X/S}$ is called the exterior differential. If X is locally of finite type over S, then $\Omega^1_{X/S}$ is of finite type by (1.11).

Let X and Y be S-schemes If $f : X \longrightarrow Y$ an S-morphism, then there exists a canonical exact sequence of O_X-Modules

$$f^*\Omega^1_{Y/S} \longrightarrow \Omega^1_{X/S} \longrightarrow \Omega^1_{X/Y} \longrightarrow 0$$

by (1.6). If $pr_1 : X\times_S Y \longrightarrow X$ and $pr_2 : X\times_S Y \longrightarrow Y$ are the projections, then

$$pr_1^*\Omega^1_{X/S} \oplus pr_2^*\Omega^1_{Y/S} = \Omega^1_{X\times_S Y/S}$$

by (1.12). Further, by (1.19) the canonical sequence

$$0 \longrightarrow pr_1^*\Omega^1_{X/S} \longrightarrow \Omega^1_{X\times_S Y/S} \longrightarrow pr_2^*\Omega^1_{Y/S} \longrightarrow 0$$

is exact and split. Finally, by (1.17),

$$\Omega^1_{X/S} \oplus \Omega^1_{Y/S} = \Omega^1_{X \amalg Y/S}.$$

Let $i : X \hookrightarrow Y$ be an immersion of S-schemes. Then, by (1.8), the sequence of O_X-Modules

$$J/J^2 \xrightarrow{\ \delta\ } i^*\Omega^1_{Y/S} \longrightarrow \Omega^1_{X/S} \longrightarrow 0$$

is exact, where J is a sheaf of ideals defining X in some neighborhood and δ is induced by $d_{Y/S}$. The O_X-Module J/J^2 is called the conormal sheaf of X in Y and is denoted $\check{N}(i)$.

If X is an S-scheme, then the diagonal morphism $\Delta_{X/S} : X \longrightarrow X\times_S X$ is an immersion. Let $J_{X/S}$ be a corresponding sheaf of ideals. Then, by (1.13),

$$\Omega^1_{X/S} = \Delta^*_{X/S}(J_{X/S}/J^2_{X/S}) = \check{N}(\Delta_{X/S}).$$

Finally, let $S' \longrightarrow S$ be a morphism, X an S-scheme, $X' = X\times_S S'$, and $f : X' \longrightarrow X$ the projection. Then, by (1.18), the canonical map

$$f^*\Omega^1_{X/S} \longrightarrow \Omega^1_{X'/S'}$$

is an isomorphism.

2. Quasi-finite morphisms

Definition (2.1). - Let X and Y be schemes and $f : X \longrightarrow Y$ a morphism locally of finite type. Then f is said to be *quasi-finite* if, for each point $x \in X$, O_x is a *quasi-finite* O_y-module, i.e., if O_x/m_yO_x is a finite dimensional vector space over the field $k(y)$.

Remark (2.2). - A finite morphism is quasi-finite.

Proposition (2.3). - Let X and Y be schemes and $f : X \longrightarrow Y$ a morphism locally of finite type. Let x be a point of X and $y = f(x)$. Then the following conditions are equivalent:

(i) O_x is a quasi-finite O_y-module.

(ii) x is isolated in its fiber; i.e., $\{x\}$ is open in $f^{-1}(f(x))$.

(iii) The following two conditions hold:

(a) There exists a positive integer r such that $m_x^r \subset m_yO_x$.

(b) The field $k(x)$ is a finite algebraic extension of $k(y)$.

Proof. We may assume that Y and X are affine with rings O_y and A and that A is an O_y-algebra of finite type. Then $f^{-1}(y) = \mathrm{Spec}(B)$ where $B = A/m_yA$. Let I be the kernel of the localization map $B \longrightarrow O_x/m_yO_x$ Since I is finitely generated, there exists $s \notin m_x/m_yB$ such that $I_s = 0$; replacing B by B_s, we may assume $B \longrightarrow O_x/m_yO_x$ is injective.

Assume (i). Then B is a finite dimensional $k(y)$-vector space; hence, by (II,4.5), B is artinian. So, by (II,4.7), $f^{-1}(y)$ is discrete and (ii) holds. Further, by (II,4.7), $(m_x/m_yO_x)^r = 0$; hence, (iii) (a) holds. Since $k(x)$ is a quotient of O_x/m_yO_x, (iii) (b) holds.

Assume (ii) holds. Replacing X by a suitable neighborhood of x, we may assume $f^{-1}(y) = \{x\}$. Then $B = O_x$; so, O_x/m_yO_x is of finite type over $k(y)$ and has only one prime ideal. Hence, by (II, 4.7), (i) holds. Finally, by (II,4.6) applied to O_x/m_yO_x, (iii) implies (i).

<u>Proposition (2.4)</u>. - Let X and Y be locally noetherian schemes, $f : X \longrightarrow Y$ a morphism locally of finite type, x a point of X and $y = f(x)$. Then O_x is quasi-finite over O_y if and only if $\hat{O}_x$ is finite over $\hat{O}_y$

<u>Proof</u>. If O_x is quasi-finite over O_y, then there exists a surjection $\varphi' : k(y)^n \longrightarrow O_x/m_yO_x$ for some integer $n > 0$; lift φ' to a map $\varphi : O_y^n \longrightarrow O_x$. By (2.3), there exists an integer $r > 0$ such that $m_x^r \subset m_yO_x$. Hence, it follows from (II,1.19 and 1.20 (ii)) that $\hat{\varphi} : \hat{O}_y^n \longrightarrow \hat{O}_x$ is surjective.

Conversely, assume there exists a surjection $\alpha : \hat{O}_y^n \longrightarrow \hat{O}_x$ for some $n > 0$. Then, by (II,1.19), α induces a surjection $k(y)^n \longrightarrow k(x)$. In view of (2.3), $\hat{m}_x^r \subset \hat{m}_y\hat{O}_x$ for some r and we are reduced to proving the following lemma.

<u>Lemma (2.5)</u>. - Let $A \longrightarrow B$ be a local homomorphism of noetherian local rings and m, n the maximal ideals. Suppose that $\hat{n}^r \subset \hat{m}\hat{B}$. Then $n^r \subset mB$.

Proof. Consider the map $\beta : n^r \longrightarrow B/mB$; by (II,1.19), β induces a map $\hat{\beta} : \hat{n}^r \longrightarrow \hat{B}/\hat{m}\hat{B}$. By hypothesis, $\hat{\beta} = 0$; hence, by (II,1.15), $\beta = 0$. Thus, $n^r \subset mB$.

3. Unramified morphisms

Definition (3.1). - Let X and Y be locally noetherian schemes, $f : X \longrightarrow Y$ a morphism locally of finite type, x a point of X and $y = f(x)$. Then f (resp. O_x/O_y) is said to be unramified at x if $m_x = m_y O_x$ and $k(x)$ is a finite separable field extension of $k(y)$, (i.e., if $O_x/m_y O_x$ is a finite separable field extension of $k(y)$).

Lemma (3.2). - Let k be a field, K an artinian k-algebra of finite type and $\bar{k}$ the algebraic closure of k. If $K \otimes_k \bar{k}$ is reduced, (i.e., without nilpotents), then K is a finite product of finite separable field extensions of k.

Proof. By (II,4.9), $K = \Pi K_i$ where K_i are artinian local rings. Replacing K by K_i, we may assume K is local. Since the maximal ideal of K is nilpotent, it is zero and thus K is a field which is finite over k by (II,4.7).

Let α be an element of K and $f(T)$ its minimal polynomial over k. Then $k(\alpha) \cong k[T]/f(T)$; so, $k(\alpha)\otimes_k \bar{k} \cong \Pi \bar{k}[T]/f_i(T)^{r_i}$ where the $f_i(T)$ are the distinct linear factors of $f(T)$. By hypothesis, $k(\alpha)\otimes_k \bar{k}$ is reduced. Hence, all $r_i = 1$; so, α is separable.

Proposition (3.3). - Let X and Y be locally noetherian schemes, $f : X \longrightarrow Y$ a morphism locally of finite type and x a point of X. Then the following conditions are equivalent:

(i) $\Omega^1_{X/Y}$ is zero at x

(ii) $\Delta_{X/Y}$ is an open immersion in neighborhood of x.

(iii) f is unramified at x.

<u>Proof</u>. Assume (i) holds. Let J be the sheaf of ideals defining the diagonal in a neighborhood of itself and identify x with $\Delta_{X/Y}(x)$. Then, by (1.13), $0 = (\Omega^1_{X/Y})_x = (J/J^2)_x$. Hence, by Nakayama's lemma, $J_x = 0$ and (ii) holds.

Assume (ii). To prove (iii), we may assume that $Y = \operatorname{Spec}(k(y))$, $f^{-1}(y) = X = \operatorname{Spec}(A)$ and that $\Delta_{X/Y} : X \longrightarrow X \times_Y X$ is an open immersion. Let k be the algebraic closure of $k(y)$. If $A' = A \otimes_{k(y)} k$ is proved isomorphic to a finite product Πk, then A will be finite dimensional over $k(y)$ and (iii) will result from (3.2).

Replace Y by $\operatorname{Spec}(k)$ and X by $X \otimes_Y k$. Let z be a closed point of X. Then by (III,2.8), $O_z/m_z \cong k$.

Consider the morphism $g = (id_X, h_z) : X \longrightarrow X \times_Y X$, where $h_z : X \longrightarrow X$ is the constant morphism through z, (defined by the composition $A \longrightarrow k(z) \xrightarrow{\sim} k \hookrightarrow A$). Then, since the diagonal subset is open, $g^{-1}(\Delta) = \{z\}$ is open. Thus, all closed points of X are open; so, all primes of A are maximal. Hence, by (II,4.7), A is artinian and X consists of a finite number of points. Then, by choosing X small enough, we may assume X consists of a single point and $A = O_x$. Since $\Delta_{X/Y}$ is an open immersion, $A \otimes_k A \longrightarrow A$ is an isomorphism. Hence, $\dim_k(A) = 1$ and $A = k$.

Assume (iii). To prove (i), we may assume $Y = \operatorname{Spec}(k(y))$ $X = f^{-1}(y)$ in view of (1.18). By (2.3) x is isolated in X. Hence, we may assume $X = \operatorname{Spec}(k(x))$. Thus, we are reduced to proving the following lemma.

<u>Lemma (3.4)</u>. - If L is a finite separable field extension of K, then $\Omega^1_{L/K} = 0$.

Proof. Let $D : L \longrightarrow M$ be a K-derivation. Let $a \in L$ and $f(T)$ be the minimal polynomial of a over K. Then $f(a) = 0$; hence, $f'(a)D(a) = 0$. Since a is separable over K, $f'(a) \neq 0$. Therefore, $D(a) = 0$.

Proposition (3.5) (Le sorite for unramified morphisms). -

(i) Any immersion is unramified.

(ii) The composition of unramified morphisms is unramified.

(iii) Any base extension of an unramified morphism is unramified.

Consequently,

(iv) The product of unramified morphisms is unramified.

(v) If $g \circ f$ is unramified, then f is unramified.

(vi) If f is unramified, then f_{red} is unramified.

Proof. Assertions (i) and (ii) are immediate from the definition. Assertion (iii) follows from (3.3 (i)) and (1.18).

Proposition (3.6). - Let X and Y be locally noetherian S-schemes and $f : X \longrightarrow Y$ an S-morphism locally of finite type. Let x be a point of X and s its projection on S. Then:

(i) f is unramified at x if and only if the canonical map $f^*\Omega^1_{Y/S} \longrightarrow \Omega^1_{X/S}$ is surjective at x.

(ii) f is unramified at x if and only if $f \otimes_S k(s) : X \otimes_S k(s) \longrightarrow Y \otimes_S k(s)$ is unramified at x.

Proof. Since the sequence $f^*\Omega^1_{Y/S} \longrightarrow \Omega^1_{X/S} \longrightarrow \Omega^1_{X/Y} \longrightarrow 0$ is exact by (1.6), (i) results from (3.3). Assertion (ii) follows immediately from the definition.

Proposition (3.7). - Let X and Y be locally noetherian schemes, $f : X \longrightarrow Y$ a morphism locally of finite type, x a point of X and $y = f(x)$. Then f is unramified at x, if and only if

$\hat{O}_x/\hat{O}_y$ is unramified. Further, suppose that $k(x) = k(y)$ or that $k(y)$ is algebraically closed. If f is unramified at x, then $\hat{O}_y \longrightarrow \hat{O}_x$ is surjective.

Proof. Assume $\hat{O}_x/\hat{O}_y$ is unramified. Then $\hat{m}_x = \hat{m}_y\hat{O}_x$. By (2.5), $m_x \subset m_yO_x$; hence, $m_x = m_yO_x$. By (II,1.19), $k(x)/k(y)$ is separable; thus, f is unramified at x. Conversely, if f is unramified at x, then, by (II,1.19), $\hat{O}_x/\hat{O}_y$ is unramified. If, further, $k(y)$ is algebraically closed, then, since $k(x)/k(y)$ is finite, $k(x) = k(y)$. Therefore, in either case, $k(y) \longrightarrow k(x)$ is bijective. Hence, by (II,1.20), $\hat{O}_y \longrightarrow \hat{O}_x$ is surjective.

4. Étale morphisms

Definition (4.1). - Let X and Y be locally noetherian schemes and $f : X \longrightarrow Y$ a morphism locally of finite type. Then f (resp. $\varphi : O_y \longrightarrow O_x$, O_x/O_y) is said to be étale at $x \in X$ if f is flat and unramified at x.

Example (4.2). - Let k be a field and $f : X \longrightarrow \mathrm{Spec}(k)$ an étale morphism. Then $X = \coprod_{i=1}^{n} \mathrm{Spec}(k_i)$ where the k_i are finite separable extensions of k.

Proof. By (2.3), X is an artinian scheme; hence, since f is unramified, O_x is a finite separable field extension of k for each $x \in X$ and $X = \coprod \mathrm{Spec}(O_x)$.

Proposition (4.3). - Let X and Y be locally noetherian schemes and $f : X \longrightarrow Y$ a morphism locally of finite type. Then f is étale at $x \in X$ if and only if $\hat{O}_x$ is étale over $\hat{O}_{f(x)}$.

Proof. The assertion holds with "étale" replaced by "flat" (V,3.3) or by "unramified" (3.7).

<u>Proposition (4.4)</u>. - Let X and Y be locally noetherian schemes and $f : X \longrightarrow Y$ a morphism locally of finite type. Suppose f is flat and quasi-finite at $x \in X$. Then $\hat{\varphi} : \hat{O}_{f(x)} \longrightarrow \hat{O}_x$ is injective and finite.

<u>Proof</u>. By (V,3.3) and (2.4), $\hat{\varphi}$ is flat and finite; whence, by (V,1.6) and (V,1.9), $\hat{\varphi}$ is injective.

<u>Corollary (4.5)</u>. - Let X and Y be locally noetherian schemes, $f : X \longrightarrow Y$ a morphism locally of finite type, x a point of X and $y = f(x)$. If $\hat{\varphi} : \hat{O}_y \longrightarrow \hat{O}_x$ is an isomorphism, then f is étale at x. Conversely, suppose that the residue extension $k(x)/k(y)$ is trivial or that $k(y)$ is algebraically closed. If f is étale at x, then $\hat{\varphi}$ is an isomorphism.

<u>Proof</u>. By (4.3), if $\hat{\varphi}$ is an isomorphism, then O_x is étale over O_y. Conversely, if f is étale at x, then $\hat{\varphi}$ is injective by (4.4) and surjective by (3.7).

<u>Proposition (4.6)</u>. - Let X and Y be locally noetherian schemes and $f : X \longrightarrow Y$ a morphism locally of finite type. If f is étale at $x \in X$, then f is étale in a neighborhood of x.

<u>Proof</u>. The assertion holds with "étale" replaced by "flat" (V,5.5) or by "unramified" (3.3).

<u>Proposition (4.7) (Le sorite for étale morphisms)</u>. -

(i) An open immersion is étale.

(ii) The composition of étale morphisms is étale.

(iii) Any base extension of an étale morphism is étale.

(iv) The product of étale morphisms is étale.

(v) If $g \circ f$ is étale and if g is unramified, then f is étale.

__Proof__. Assertions (i), (ii), (iii), and (iv) each hold with "étale" replaced by "flat" (V,2.7) or by "unramified" (3.5). As to (v), consider the diagram with cartesian squares:

$$\begin{array}{ccccc} Y & \xrightarrow{\Delta_{Y/S}} & Y\times_S Y & & \\ {\scriptstyle f}\uparrow & \square & \uparrow{\scriptstyle f\times id_Y} & & \\ X & \xrightarrow[\Gamma_f]{} & X\times_S Y & \xrightarrow{pr_2} & Y \\ & & \downarrow & \square & \big| \\ & & X & \xrightarrow{g\circ f} & S \end{array}$$

Since $g \circ f$ is étale, pr_2 is étale by (iii). Since $\Delta_{Y/S}$ is an open immersion by (3.3), Γ_f is étale by (i) and (iii). Therefore, $f = pr_2 \circ \Gamma_f$ is étale by (ii).

__Proposition (4.8)__. - Let S be a locally noetherian scheme, X and Y two schemes locally of finite type over S and $f : X \longrightarrow Y$ an S-morphism. Let x be a point of X and s its image in S. Suppose X and Y are flat over S. Then f is flat (resp. étale) at x if and only if $f_s = f \otimes_S k(s)$ is flat (resp. étale) at x.

__Proof__. The first assertion follows from (V,3.4); the second, from the first and (3.6(ii)).

__Proposition (4.9)__. - Let S be a locally noetherian scheme, X and Y two schemes locally of finite type over S and $f : X \longrightarrow Y$ an S-morphism. If f is étale, then the canonical map

$$f^*\Omega^1_{Y/S} \longrightarrow \Omega^1_{X/S}$$

is an isomorphism.

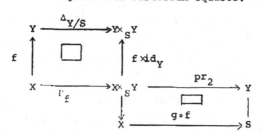

Proof. Consider the diagram with cartesian square

$$\begin{array}{ccccc}
X & \xrightarrow{\Delta_{X/Y}} & X\times_Y X & \xrightarrow{g'} & X\times_S X \\
 & {\scriptstyle f}\searrow & \downarrow{\scriptstyle h'} & \square & \downarrow{\scriptstyle h=f\times_S f} \\
 & & Y & \xrightarrow[\Delta_{Y/S}]{} & Y\times_S Y
\end{array}$$

By (3.3), $\Delta_{X/Y}$ is an open immersion. Hence, by (1.21), $\Omega^1_{X/S} =$ $= \check{N}(g'\circ \Delta_{X/Y}) = \Delta^*_{X/Y}(\check{N}(g'))$. By the lemma below, $\check{N}(g') = h'^*(\check{N}(\Delta_{Y/S}))$ and, by (1.21), $\check{N}(\Delta_{Y/S}) \cong \Omega^1_{Y/S}$; whence, the assertion.

Lemma (4.10). - Consider a cartesian diagram

$$\begin{array}{ccc}
X' & \xhookrightarrow{g'} & Y' \\
{\scriptstyle h'}\downarrow & & \downarrow{\scriptstyle h} \\
X & \xhookrightarrow[g]{} & Y
\end{array}$$

where g and g' are immersions of schemes. If h is flat, then the induced map on conormal sheaves $h'^*\check{N}(g) \longrightarrow \check{N}(g')$ is an isomorphism.

Proof. Let J be the quasi-coherent sheaf of ideals defining X as a subscheme of Y in a neighborhood U of X. Since h is flat, the sequence

$$0 \longrightarrow J\otimes_{O_Y} O_{Y'} \longrightarrow O_{Y'} \longrightarrow O_{X'} \longrightarrow 0$$

is exact; hence, $J' = J\otimes_{O_Y} O_{Y'}$ is the ideal defining X' in $h^{-1}(U)$. Therefore the diagram

$$\begin{array}{ccccccc}
(J\otimes_{O_Y}O_{Y'})\otimes_{O_{Y'}}(J\otimes_{O_Y}O_{Y'}) & \longrightarrow & J\otimes_{O_Y}O_{Y'} & \longrightarrow & \check{N}(g)\otimes_{O_Y}O_{Y'} & \longrightarrow & 0 \\
\downarrow\wr & & \downarrow\wr & & \downarrow & & \\
J'\otimes_{O_{Y'}}J' & \longrightarrow & J' & \longrightarrow & \check{N}(g') & \longrightarrow & 0
\end{array}$$

yields the assertion.

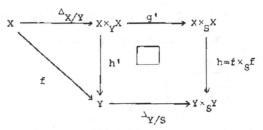

5. Radicial morphisms

Definition (5.1). - A morphism $f : X \longrightarrow Y$ of schemes is said to be radicial if it is injective and if, for all $x \in X$, the residue extension $k(x)/k(f(x))$ is purely inseparable (radicial).

Proposition (5.2). - Let $f : X \longrightarrow Y$ be a morphism of schemes. The following conditions are equivalent:

(i) f is radicial.

(ii) For any field K, the map of K-points $f(K) : X(K) \longrightarrow Y(K)$ is injective.

(iii) (Universal injectivity) For any base extension $Y' \longrightarrow Y$, the morphism $f_{Y'}: X \times_Y Y' \longrightarrow Y'$ is injective.

(iv) (Geometric injectivity) For any field K and any morphism $\mathrm{Spec}(K) \longrightarrow Y$, the morphism $f_K: X \otimes_Y K \longrightarrow \mathrm{Spec}(K)$ is injective.

Proof. Assume (i) and for some field K, let $u_1, u_2: \mathrm{Spec}(K) \rightrightarrows X$ satisfy $f \circ u_1 = f \circ u_2$. Since f is injective, $x = \mathrm{Im}(u_1) = \mathrm{Im}(u_2)$. Hence, u_1, u_2 corresponds to $k(f(x))$-homomorphisms $k(x) \rightrightarrows K$. Since $k(x)/k(f(x))$ is purely inseparable, $u_1 = u_2$ and (ii) holds.

Conversely, assume (ii) and suppose $k(x)/k(f(x))$ were not purely inseparable for some $x \in X$. Then there would exist two different $k(f(x))$-homomorphisms of $k(x)$ into some field K. Let $u_1, u_2 : \mathrm{Spec}(K) \rightrightarrows X$ be the corresponding morphisms. Then $f \circ u_1 = f \circ u_2$, but $u_1 \neq u_2$.

Suppose $f(x_1) = f(x_2) = y$ for distinct points $x_1, x_2 \in X$. Then there exists a field K and two $k(y)$-homomorphisms $k(x_1) \longrightarrow K$ and $k(x_2) \longrightarrow K$. Let $u_1, u_2: \mathrm{Spec}(K) \rightrightarrows X$ be the corresponding morphisms. Then $f \circ u_1 = f \circ u_2$, but $u_1 \neq u_2$. Therefore (i) holds.

Assume (ii). Then the diagram

$$\begin{array}{ccc} (X\times_Y Y')(K) & = & X(K)\times_{Y(K)} Y'(K) \\ \downarrow & & \downarrow \\ Y'(K) & = & Y(K)\times_{Y(K)} Y'(K) \end{array}$$

shows that $f_{Y'}$ also satisfies (ii). So, by (ii) $\Longrightarrow$ (i), $f_{Y'}$ is injective and (iii) holds. The implication (iii) $\Longrightarrow$ (iv) is trivial.

Assume (iv) and, for some field K, let $u_1, u_2 \in X(K)$ satisfy $f\circ u_1 = f\circ u_2$. Then u_1 and u_2 give rise to sections $u_1', u_2' : \mathrm{Spec}(K) \rightrightarrows X\otimes_Y K$.

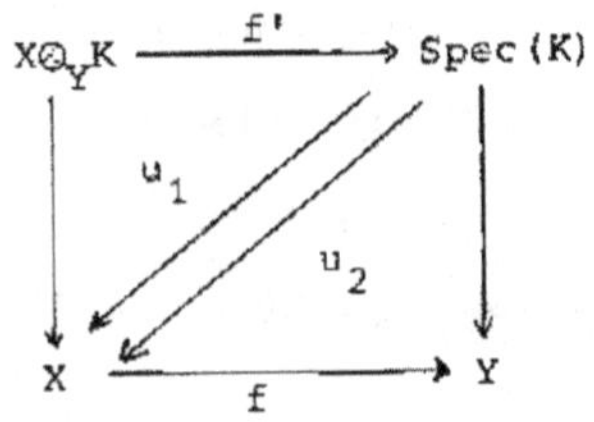

Since f' is injective, $X\otimes_Y K$ consists of a single point. It follows that $u_1' = u_2'$, so $u_1 = u_2$. Thus, (ii) holds and the proof is complete.

<u>Proposition (5.3) (Le sorite for radicial morphisms)</u>. -

(i) Any immersion, (in fact, any monomorphism), is radicial.

(ii) The composition of radicial morphisms is radicial.

(iii) Any base extension of a radicial morphism is radicial.

Consequently,

(iv) The product of radicial morphisms is radicial.

(v) If $g\circ f$ is radicial, then f is radicial.

(vi) If f is radicial, then f_{red} is radicial.

<u>Proof</u>. Assertions (i), (ii) and (iii) follow immediately from (5.2).

<u>Lemma (5.4)</u>.- Let B be a noetherian ring and S a multiplicative subset. Suppose the canonical map $B \longrightarrow S^{-1}B$ is surjective. Then for a suitable ring, C, the rings B and $S^{-1}B \times C$ are isomorphic.

<u>Proof.</u> Since the kernel I of $B \longrightarrow S^{-1}B$ is finitely generated, there is an $s \in S$ such that $sI = 0$. Therefore, $U = \mathrm{Spec}(S^{-1}B)$ is an open subscheme of $X = \mathrm{Spec}(B)$. Since $B \longrightarrow S^{-1}B$ is surjective, U is also closed.

It follows that there exists a ring C such that the open subscheme $X - U$ is equal to $\mathrm{Spec}(C)$. Then, $B = S^{-1}B \times C$.

<u>Theorem (5.5)</u>. - Let X and Y be locally noetherian schemes. Then a morphism $f : X \longrightarrow Y$ is an open immersion if (and only if) f is étale and radicial.

<u>Proof</u>. Since f is flat, it is open by (V,5.1). Since f is also injective, it is a homeomorphism onto its image. It remains to show that, for each $x \in X$, the map $O_{f(x)} \longrightarrow O_x$ is an isomorphism. Set $A = \hat{O}_{f(x)}$ and $B = O_x \otimes_{O_{f(x)}} \hat{O}_{f(x)}$. Since A is faithfully flat over $O_{f(x)}$, it suffices to show that $A \longrightarrow B$ is an isomorphism

Let m be the maximal ideal of A and n a maximal ideal of B containing mB. Then $A \longrightarrow B_n$ is a local homomorphism and is étale and radicial by (4.7) and (5.3). Since the residue extension of B_n over A is both separable and purely inseparable, it is trivial. Consider the commutative diagram

$$\begin{array}{ccc} A & \longrightarrow & B_n \\ \| & & \downarrow \\ \hat{A} & \longrightarrow & \hat{B}_n \end{array}$$

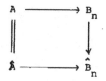

The map $\hat{A} \longrightarrow \hat{B}_n$ is an isomorphism by (4.5) and $B_n \longrightarrow \hat{B}_n$ is injective by (II,1.15). Hence, $A \xrightarrow{\sim} B_n \xrightarrow{\sim} \hat{B}_n$ and $B \longrightarrow B_n$ is surjective. Therefore, by (5.4), for a suitable ring C, $B \cong A \times C$. However, B is radicial over A, so $\text{Spec}(B) \longrightarrow \text{Spec}(A)$ is injective; hence, $\text{Spec}(C) = \emptyset$, $C = 0$, and $A \xrightarrow{\sim} B$.

Corollary (5.6). An étale monomorphism is an open immersion.

Proposition (5.7). - Let S be a locally noetherian scheme, X and Y two schemes locally of finite type over S and $f : X \longrightarrow Y$ an S-morphism. Suppose X and Y are flat over S. Then, f is an open immersion if and only if $f \otimes_S k(s) : X \otimes_S k(s) \longrightarrow Y \otimes_S k(s)$ is an open immersion for all $s \in S$.

Proof. The assertion holds with "open immersion" replaced by "étale morphism" (4.8) and by "radicial morphism" (5.1); hence, the assertion follows from (5.5).

6. Covers

Definition (6.1). - Let X, Y be locally noetherian schemes and $f : X \longrightarrow Y$ a morphism locally of finite type. Then X is said to be a (ramified) cover of Y (resp. f is said to be a covering (map)) if f is finite and surjective; X is said to be an unramified (resp. flat, étale) cover of Y if, further, f is unramified (resp. flat, étale).

Proposition (6.2). - Let X, Y be locally noetherian schemes. If X is a cover of Y, then $\dim(X) = \dim(Y)$.

Proof. It is clear that $\dim(X) = \sup\{\dim(O_x)\}$. Hence, replacing Y by an open subset U (resp X by $f^{-1}(U)$), we may assume that Y (resp. X) is affine with ring A (resp. B) and that B is a finite A-module. Then, it follows by induction from (III,2.2) that $\dim(B) = \dim(A)$.

Definition (6.3).- Let X, Y be locally noetherian schemes and $f : X \longrightarrow Y$ a morphism locally of finite type. The set of points of X where f is ramified is called the branch locus of X over Y.

Remark (6.4). - The branch locus of X over Y has a natural, closed subscheme structure defined by the annihilator $\vartheta_{X/Y}$ of $\Omega^1_{X/Y}$; $\vartheta_{X/Y}$ is often called the Kähler different of X over Y.

Remark (6.5). - Let A be a ring, E a finite, free A-module and $h : E \longrightarrow E$ an A-homomorphism. If $M(h)$ is the matrix of h with respect to some basis, then the trace of h, denoted $tr(h)$, is defined as the sum of the diagonal elements of $M(h)$ and is clearly independent of the choice of basis If $\varphi : A \longrightarrow B$ is a ring homomorphism, then $E \otimes_A B$ is a free B-module, $h \otimes id_B : E \otimes_A B \longrightarrow E \otimes_A B$ is a B-homomorphism and $tr(h \otimes id_B) = \varphi(tr(h))$.

Let X be a cover of Y and F a coherent O_X-Module, flat over Y. Then the trace of an endomorphism g of F may be defined. Namely, by (V,2.8), there exists an open affine cover V_α of Y such that $f_*F|_{V_\alpha}$ is free and the elements $tr(g|_{f^{-1}(V_\alpha)}) \in \Gamma(V_\alpha, O_Y)$ piece together to give an element $tr(g) \in \Gamma(Y, O_Y)$. Furthermore, a map $Tr : \underline{End}_{O_Y}(f_*F) \longrightarrow O_Y$ exists where $Tr_V(g)$ is the trace of $g|_V$. In particular, if X is a flat cover of Y, then $Tr_{X/Y} : f_*O_X \longrightarrow O_Y$ is defined as the composition of the canonical map $f_*O_X \longrightarrow \underline{End}_{O_Y}(f_*O_X)$ with Tr.

There exists a map associated to $Tr_{X/Y}$,

$$u = astr_{X/Y} : f_*O_X \longrightarrow (f_*O_X)^\vee = \underline{Hom}_{O_Y}(f_*O_X, O_Y),$$

defined as follows: For an open set V of Y and elements $a, b \in \Gamma(f^{-1}(V), O_X)$, let $u_V(a)$ be the map taking b to

$(Tr_{X/Y})_V(ab) \in \Gamma(V,O_Y)$. Let $\Lambda^{max} f_* O_X$ denote the invertible sheaf equal to $\Lambda^r f_* O_X$ where $f_* O_X$ has rank r. Then the section $\Lambda^{max} u \in Hom(\Lambda^{max} f_* O_X, \Lambda^{max}(f_* O_X)^\vee)$ is called the <u>discriminant</u> and is denoted $d_{X/Y}$. The image of $d_{X/Y} \otimes id : \Lambda^{max} f_* O_X \otimes \Lambda^{max} f_* O_X \longrightarrow O_Y$ is called the <u>discriminant ideal</u> and denoted $D_{X/Y}$. The set of points of Y where $D_{X/Y}$ is not equal to O_Y is called the <u>discriminant locus</u>.

<u>Proposition (6.6)</u>. - Let X, Y be noetherian affine schemes with rings B, A and suppose B is a finite, free A-module. Then the following conditions are equivalent:

(i) X is an étale cover of Y.

(ii) The pairing $(a,b) \longmapsto tr_{B/A}(ab)$ is nonsingular.

(iii) The discriminant ideal $D_{X/Y}$ is equal to A.

<u>Proof</u>. The equivalence (ii) $\Longleftrightarrow$ (iii) follows easily from the definitions. Since X is a flat cover, it is étale if and only if it is unramified; hence, by (3.6), if and only if for every $y \in Y$, the n-dimensional $k(y)$-algebra $B \otimes_A k(y)$ is separable (unramified over $k(y)$). Let k be the algebraic closure of $k(y)$. By (6.5), the trace commutes with the base extension $A \longrightarrow k$; so, we may assume $A = k$ and, by (II,4.9), $B = \Pi_{i=1}^r B_i$ where the B_i are artinian local rings. Since $tr_{B/A} = \Sigma tr_{B_i/A}$, we may assume $r = 1$; then, by (3.2),(3.3) and (3.4), it remains to show that the pairing is nonsingular if and only if B is a field.

Let m be the maximal ideal of B. By (II,4.7), there exists an s such that $m^s = 0$, but $m^{s-1} \neq 0$. If $s = 1$, then $B = k$ and $tr_{B/A}(ab) = ab$ is clearly nonsingular. If $s \geq 2$, then since $B = k \oplus m$, it follows that $tr_{B/A}(y) = 0$ for all $y \in m^{s-1}$. Let x be a nonzero element of m^{s-1}. Since $xb \in m^{s-1}$ for all $b \in B$,

$tr_{B/A}(xb) = 0$ for all $b \in B$; so, the pairing is singular.

Lemma (6.7). - Let B be a semilocal ring and $m_1,\ldots,m_r$ the maximal ideals of B. Then $\hat{B} = \Pi \hat{B}_{m_i}$.

Proof. Let q be an ideal of definition. It follows from (II,4.9) that $B/q^r = \Pi(B/q^r)_{m_i} = \Pi B_{m_i}/q^r B_{m_i}$. Therefore, by (II,1.8), $\hat{B} = \Pi \hat{B}_{m_i}$.

Theorem (6.8) (Purity of the branch locus). - Let X and Y be locally noetherian schemes. If X is a flat cover of Y, then the branch locus of X and Y has pure codimension 1.

Proof. Let x be a ramified point of X and $y = f(x)$. It suffices to show that ϑ_{O_x/O_y} is contained in a height 1 prime of O_x. Let B be the affine ring of $X\times_Y \mathrm{Spec}(O_y)$. Then B is a finite O_y-module; hence, a semilocal ring with radical $m_y B$ (2.3). By (II,1.18) and (6.7), $B\otimes_{O_y}\hat{O}_y = \hat{B} = \Pi \hat{O}_{x_i}$ where x_i runs through the points of $f^{-1}(y)$ and, by (1.18) and (1.17), $\vartheta_{B/O_y}\hat{O}_y = \vartheta_{\hat{B}/\hat{O}_y} = \\ = \oplus \vartheta_{\hat{O}_{x_i}/\hat{O}_y}$; hence, $\vartheta_{\hat{O}_x/\hat{O}_y} = \vartheta_{O_x/O_y}\hat{O}_y$. Therefore, by (V,3.3) and (III,1.8), we may assume $O_x = \hat{O}_x$ and $O_y = \hat{O}_y$.

By (2.4), O_x is a flat cover of O_y; but, by (3.7), not étale. Hence, by (6.6), $D_{O_x/O_y} \subset m_y$. Therefore, by (III,1.10 and 5.8), $D_{O_x/O_y} \subset p$ where p is a height 1 prime of O_y. By (6.6), O_x is not étale over O_y at some prime q of O_x lying over p. By (V,2.10), q has height 1; whence, the assertion.

Lemma (6.9). - Let A be a ring, B an A-algebra and let $t \in B$ generate B over A. If $P \in A[T]$ is a polynomial such that $P(t) = 0$, then $\vartheta_{B/A} \supset P'(t)B$ where $P'(T) = \frac{d}{dT}P(T)$; furthermore, if the natural map $A[T]/PA[T] \longrightarrow B$ is an isomorphism, then $\vartheta_{B/A} = P'(t)B$.

Proof. The canonical map $A[T] \longrightarrow B$ is surjective; let I be its kernel. By (1.8), the sequence

$$I/I^2 \longrightarrow \Omega^1_{A[T]/A} \otimes_A B \longrightarrow \Omega^1_{B/A} \longrightarrow 0$$

is exact. Since, by (1.4), $\Omega^1_{A[T]/A} \otimes_A B = BdT$, it follows that $\Omega^1_{B/A} = B/d(I)B$ where $d(I) = \{\frac{d}{dT} Q(T) | Q(T) \in I\}$. Hence, $d(I)B = \vartheta_{B/A}$. Thus, $\vartheta_{B/A} \supset P'(t)B$ and, if $I = PA[T]$, then $d(I)B = P'(t)B$; so $\vartheta_{B/A} = P'(t)B$.

Proposition (6.10). - Let A be a noetherian ring, B an A-algebra, q a prime of B and p the trace of q in A. Suppose there exists a polynomial $P(T)$ and an element $t \in B$ such that the map $A[T]/PA[T] \longrightarrow B$ defined by t is an isomorphism. Then B_q is unramified over A_p if and only if $(P,P')A_p[T] = A_p[T]$. Suppose, in addition, that the leading coefficient of P is invertible. Then B_q is étale over A_p if and only if $P'(t) \notin q$.

Proof. Since, by (6.9), $\vartheta_{B/A} = P'(t)B$, it follows by (6.4) that B_q is unramified over A_p if and only if $P'(t)$ is a unit in B_q; hence, if and only if $(P,P')A_p[T] = A_p[T]$. The second assertion follows from the first since, if the leading coefficient of P is invertible, then B is the free A-module generated by $1,t,\ldots,t^{n-1}$ where $n = \deg(P)$.

Definition (6.11). - Let A be a ring. A polynomial $P \in A[T]$ is said to be separable if it satisfies the following two conditions:

(a) The leading coefficient of P is a unit in A.

(b) $(P,P')A[T] = A[T]$.

Theorem (6.12). - Let A be a noetherian local ring, m the maximal ideal and $k = A/m$. Let B be a finite A-algebra, $K = B \otimes_A k$ and $r = [K:k]$. Suppose either that k is infinite or that B is local. Then B is étale (resp. unramified) over A (if and) only if B is isomorphic to an étale algebra of the form A[T]/PA[T] (resp. a quotient of A[T]/PA[T]) for some separable polynomial P of degree r.

__Proof__. It follows from the hypothesis that there is a primitive element $u \in K$; say, $1,u,\ldots,u^{r-1}$ form a basis for K over k. Let $t \in B$ be an element whose residue class is u. By Nakayama's lemma, $1,t,\ldots,t^{r-1}$ generate B. If $t^r = \sum_{i=0}^{r-1} a_i t^i$, then let $P(T) = T^r - \sum a_i T^i$. From (6.10) applied K/k, it follows that $(P,P')A[T] = A[T] \bmod mA[T]$. Hence, by Nakayama's lemma, (P,P') generates A[T], so P is a separable polynomial. Finally, if B/A is étale, the assertion follows from (4.7) and (5.6) applied to the surjection $A[T]/PA[T] \longrightarrow B$.

Chapter VII - Smooth Morphisms

1. Generalities

<u>Definition (1.1)</u>. - Let X and Y be locally noetherian schemes and $f : X \longrightarrow Y$ a morphism. Then X is said to be <u>smooth over</u> Y <u>at</u> $x \in X$ (resp. f is said to be <u>smooth at</u> x) if there exists a neighborhood U of x and a commutative diagram

$$\begin{array}{ccc} U & \xrightarrow{\ g\ } & \mathbb{A}^n_Y = \mathbb{A}^n_Z \times Y \\ & {\scriptstyle f|U} \searrow & \downarrow {\scriptstyle p} \\ & & Y \end{array}$$

where g is étale and p is the projection on the second factor. (The morphism p is sometimes called a <u>polynomial morphism</u>). The scheme X is said to be <u>smooth</u> over Y (resp. f is said to be <u>smooth</u>) if f is smooth at every $x \in X$.

<u>Remark (1.2)</u>. - The points $x \in X$ at which a morphism $f : X \longrightarrow Y$ is smooth form an open set.

<u>Definition (1.3)</u>. - Let $f : X \longrightarrow Y$ be a morphism of schemes and x a point of X. The <u>relative dimension of</u> X <u>over</u> Y <u>at</u> x (resp. <u>of</u> f <u>at</u> x) is defined as the largest dimension of the components of $f^{-1}(f(x))$ passing through x and is denoted $\dim_x(X/Y)$ (resp. $\dim_x(f)$).

<u>Proposition (1.4)</u>. - In the definition of smoothness,

$$n = \dim_x(f)$$

<u>Proof</u>. Changing the base, we may assume, by (VI,4.7), that

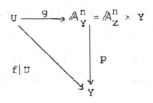

$Y = \mathrm{Spec}(k(y))$ where $y = f(x)$. Then since $\dim(\mathbb{A}^n_{k(y)}) = n$ by (III,2.6), the assertion follows from (V,2.10) and (VI,2.3) applied to g.

Remark (1.5). - If $f : X \longrightarrow Y$ is a quasi-finite morphism, then $\dim_x(f) = 0$ for all $x \in X$.

Proposition (1.6). - Let X, Y be locally noetherian schemes. A morphism $f : X \longrightarrow Y$ is étale if and only if it is smooth and quasi-finite.

Proof. As f is quasi-finite, $\dim_x(f) = 0$ by (1.5); hence, the assertion follows from the definition of smoothness and (1.4).

Proposition (1.7). (Le sorite for smooth morphisms). -

(i) An open immersion is smooth.

(ii) The composition of smooth morphisms is smooth.

(iii) Any base extension of a smooth morphism is smooth.

Consequently,

(iv) The product of smooth morphisms is smooth.

Proof.

(i) An open immersion is étale.

(ii) Since smoothness is local on X, it suffices to consider a commutative diagram with cartesian square

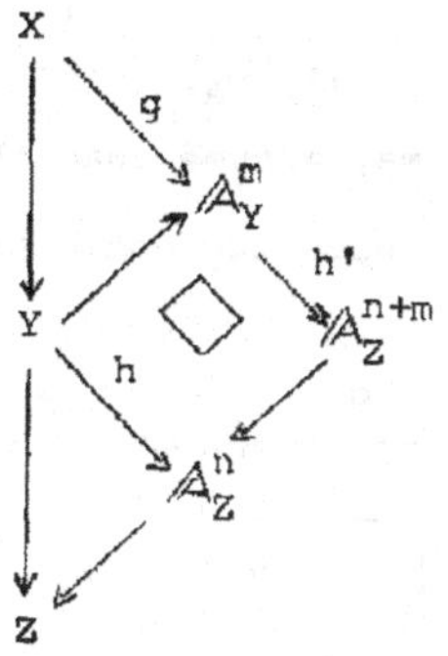

Since h' is a base extension of h, h' is étale; so, since g is étale, $h' \circ g$ is étale by (VI,4.7).

(iii) Again, it suffices to consider a commutative diagram with cartesian squares.

$$\begin{array}{ccccc} X' & \xrightarrow{g'} & \mathbb{A}^n_{Y'} & \xrightarrow{p'} & Y' \\ \downarrow & \square & \downarrow & \square & \downarrow \\ X & \xrightarrow{g} & \mathbb{A}^n_Y & \xrightarrow{p} & Y \end{array}$$

Since g is étale, it follows by (VI,4.7) that g' is étale.

<u>Theorem (1.8)</u>. - Let X, Y be locally noetherian schemes, $f : X \longrightarrow Y$ a morphism locally of finite type, x a point of X and $y = f(x)$. Then f is smooth at x if and only if the following two conditions hold:

(a) f is flat at x.

(b) $f^{-1}(y)$ is smooth over $k = k(y)$ at x.

<u>Proof</u>. If f is smooth at x, then (b) holds by (1.7). Since an étale morphism and a polynomial morphism are each flat, f is flat by (V,2.7).

To prove the converse, we may assume that X, Y are affine with rings B, A, and that there exists a factorization of f_y, $f^{-1}(y) \xrightarrow{g_y} \mathbb{A}^n_k \longrightarrow \mathrm{Spec}(k)$ where g_y is étale. If g_y is defined by n functions $g_{y,i} \in B \otimes_A k$, then replacing $g_{y,i}$ by $a g_{y,i}$ for a suitable $a \in k$, we may assume that the $g_{y,i}$ are images of functions $g_i \in B$. Then we have the commutative diagram with cartesian squares

$$\begin{array}{ccccc} X \otimes_Y k & \xrightarrow{g_y} & \mathbb{A}^n_Y \otimes_Y k & \longrightarrow & \mathrm{Spec}(k) \\ \downarrow & \square & \downarrow & \square & \downarrow \\ X & \xrightarrow{g} & \mathbb{A}^n_Y & \longrightarrow & Y \end{array}$$

where g is the morphism defined by the $g_i \in B$. Since X and $\mathbb{A}^n_Y$

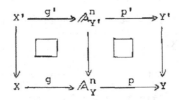

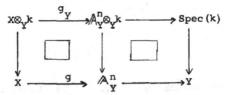

are flat over Y and g_y is étale, it follows from (VI,4.8) that g is étale.

Corollary (1.9). - Let S be a locally noetherian scheme, X,Y schemes locally of finite type over S. Let $f:X \to Y$ be an S-morphism, x a point of X with image $s \in S$. Suppose Y is flat over S. Then f is smooth at x if (and only if) the following two conditions are satisfied:

(a) X is flat over S at x.

(b) $f_s : X_s \longrightarrow Y_s$ is smooth at x.

Proof. By (VI,4.8), f is flat at x. However, $f_y = f_s \otimes_Y k(y)$; so, $f^{-1}(y)$ is smooth by (1.7) and the assertion follows from (1.8).

2. Serre's criterion

Definition (2.1). - A locally noetherian scheme X is said to satisfy condition R_k if X is regular in codimension $\leq k$ or, equivalently, if the singular locus has codimension $> k$; X is said to satisfy condition S_k if, for all $x \in X$,

$$\operatorname{depth}(O_x) \geq \inf\{k, \dim(O_x)\}.$$

A noetherian ring A is said to satisfy R_k (resp. S_k) if $X = \operatorname{Spec}(A)$ satisfies R_k (resp. S_k). A locally noetherian scheme X is said to satisfy R_k (resp. S_k) at x if O_x satisfies R_k (resp. S_k).

Proposition (2.2). - Let X be a locally noetherian scheme. Then:

(i) If $k' \geq k$, then $S_{k'}$ implies S_k and $R_{k'}$ implies R_k.

(ii) X satisfies S_k for all k if and only if X is Cohen-Macaulay.

(iii) X satisfies R_k for all k if and only if X is regular.

(iv) X satisfies S_1 if and only if X has no embedded components.

(v) X satisfies R_0 if and only if X is generically reduced (i.e., reduced in a neighborhood of each generic point).

(vi) X satisfies R_0 and S_1 if and only if X is reduced.

<u>Proof</u>. Assertions (i), (ii) and (iii) are trivial. To prove (iv), note that X satisfies S_1 if and only if $\mathrm{depth}(O_x) \geq 1$ for all $x \in X$ which are not generic points. On the other hand, $\mathrm{depth}(O_x) = 0$ if and only if $x \in \mathrm{Ass}(O_X)$ by (III,3.11). Hence, X satisfies S_1 if and only if every $x \in \mathrm{Ass}(O_X)$ is generic, i.e., if and only if X has no embedded components.

To prove (v), note that X satisfies R_0 if and only if X is generically regular and that X is generically regular if and only if X is generically reduced. Finally, to prove (vi), it suffices, in view of (iv) and (v), to prove the following lemma.

<u>Lemma (2.3)</u>. - A locally noetherian scheme X is reduced if and only if it is generically reduced and has no embedded components.

<u>Proof</u>. Since the statement is local, we may assume X is affine with ring A. Then, by the weak Nullstellensatz (II,2.8), A is reduced if and only if $0 = \cap p_i$ where the p_i are minimal primes. However, by (II,3.17), $\{p_i\}$ is an irredundant primary decomposition of 0 if and only if each A_{p_i} is reduced and all essential primes of 0 are minimal.

<u>Definition (2.4)</u>. Let A be an integral domain with quotient field K. Then A is said to be a <u>discrete</u> (rank 1) <u>valuation ring</u> if $A = \{x \in K^* | v(x) \geq 0\} \cup \{0\}$ where v is a surjective function from K^* to Z satisfying:

(i) $v(xy) = v(x) + v(y)$ for all $x,y \in K^*$.

(ii) $v(x+y) \geq \inf\{v(x),v(y)\}$ for all $x,y \in K^*$.

An element $t \in A$ is called a <u>uniformizing parameter</u> if $v(t) = 1$.

<u>Lemma (2.5)</u>. - Let A be a discrete valuation ring and t a uniformizing parameter. Then every nonzero ideal I of A is generated by t^r for some $r \geq 0$; in particular, A is a local noetherian domain.

<u>Proof</u>. Let $y \in I$ have the property that $r = v(y)$ is minimal, and let $u = y/t^r$. Then $v(u) = 0$, so u is a unit of A. Hence, $t^r = u^{-1}y \in I$. If $x \in I$, then $x = t^r x'$ where $v(x') \geq 0$. Hence, $I = t^r A$.

<u>Proposition (2.6)</u>. - Let A be a local noetherian domain with maximal ideal m. Then the following conditions are equivalent:

(i) A is a discrete valuation ring.

(ii) A is principal and is not a field

(iii) A is normal (i.e., integrally closed in its quotient field) and $\dim(A) = 1$.

(iv) A is normal and $\operatorname{depth}(A) = 1$.

(v) $m = tA$ for some nonzero $t \in A$.

<u>Proof</u>. The implication (i) $\Longrightarrow$ (ii) follows from (2.5) and (ii) $\Longrightarrow$ (iii) is easy. Since A is a domain, $\operatorname{depth}(A) \geq 1$; so, by (III,3.15), (iii) $\Longrightarrow$ (iv).

Assume (iv). Then there exists an element $x \in m$ such that $m \in \operatorname{Ass}(A/xA)$ by (III,3.10) and 3.11). Hence, there exists $y \in A$, $y \notin xA$ and such that $my \subset xA$. Then $myx^{-1} \subset A$ and $yx^{-1} \not\subset A$. It follows that $myx^{-1} = A$. For, otherwise, $myx^{-1} \subset m$ and, since m is finitely generated, yx^{-1} would be integral over A. Since A is normal, yx^{-1} would be in A. Hence, there exists $t \in m$ such that $tyx^{-1} = 1$. Now, if $z \in m$, then $t(yx^{-1}z) = z$ and $yx^{-1}z \in A$; hence (v) holds.

Assume (v). If $y \in m^r - m^{r+1}$, define $v(y) = r$. Since, by Krull's intersection theorem (II,1.15), $\cap m^r = 0$, $v(y)$ is defined for all nonzero y in A. Clearly, $v(x+y) \geq \inf\{v(x), v(y)\}$ for any $x, y \in A$. Further, since $m^r = t^r A$, if $v(y) = r$, then $y = ut^r$ for $u \in A^*$ and it follows that $v(xy) = v(x) + v(y)$; so, A is a discrete valuation ring.

<u>Proposition (2.7)</u>. - Let A be a noetherian ring which is reduced and integrally closed in its total quotient ring K. Then A is a product of normal domains.

<u>Proof</u>. By (2.3), 0 has no embedded essential primes; so, by (II,3.17 and 4.7), K is artinian. By (II,4.9), $K = \Pi K_i$ where the K_i are fields. If $e_i = (0,\ldots 0,1,0,\ldots 0)$ with 1 in the ith place, then $e_i^2 - e_i = 0$; so, since A is integrally closed, $e_i \in A$. Therefore, $A = \Pi A e_i$.

<u>Lemma (2.8)</u>. - If a local ring A has the form $A = A_1 \times \ldots \times A_r$, then $r = 1$.

<u>Proof</u>. Let m be the maximal ideal of A and $e_i = (0,\ldots,0,1,0\ldots,0)$ with 1 in the ith place. If $r > 1$, then $e_i e_j = 0$ for $i \neq j$: so, all $e_i \in m$; hence, $1 = e_1 + \ldots + e_r \in m$, a contradiction.

<u>Corollary (2.9)</u>. - A reduced noetherian local ring which is integrally closed in its total quotient ring is a normal domain.

<u>Lemma (2.10)</u>. - Let A be a noetherian ring and K its total quotient ring. If p runs through all primes such that $\operatorname{depth}(A_p) = 1$, then the sequence $A \longrightarrow K \xrightarrow{u} \Pi K_p/A_p$ is exact.

<u>Proof</u>. Let $b \in A$ be a non-zero-divisor. If p is an essential

prime of bA, then, by (II,3.9, III,3.10 and III,3.11), $\operatorname{depth}(A_p) = 1$. Thus, if $a/b \in \ker(u)$, then $a \in bA_p$ for all essential primes of bA; hence, by (II,3.17), $a \in bA$ and $a/b \in A$.

Theorem (2.11). - Let A be a noetherian ring and K the total quotient ring of A. Then the following conditions are equivalent:

(i) A satisfies R_1 and S_2.

(ii) A satisfies R_1 and S_1 and, if q runs through the primes of height 1, then the sequence $A \longrightarrow K \longrightarrow \Pi K_q/A_q$ is exact.

(iii) A is reduced and integrally closed in K.

Proof. By (2.2), A is reduced and satisfies R_0 and S_1 under all three conditions. The implication (i)$\Longrightarrow$(ii) follows from (2.10) and (2.6).

If $c \in K$ is integral over A, then its image $c_q \in K_q$ is integral over A_q for any prime q. If q has height 1, then, by R_1 and (2.6), A_q is normal; thus, $c_q \in A_q$. Hence, if (ii) holds, then $c \in A$ and (iii) holds. The implication (iii)$\Longrightarrow$(i) follows from (2.9) and (2.6).

Corollary (2.12). - Let A be a noetherian domain. Then the following conditions are equivalent:

(i) A is normal.

(ii) For all height 1 primes p, A_p is regular and the essential primes of each nonzero element have height 1.

(iii) For all height 1 primes p, A_p is a discrete valuation ring and $A = \cap A_p$ as p runs through the height 1 primes.

Corollary (2.13) (Serre's criterion). - A locally noetherian scheme X is normal if and only if it satisfies R_1 and S_2.

Proof. The assertion follows from (2.11) and (2.9).

Corollary (2.14). - Let Y be a Cohen-Macaulay scheme and X a closed subscheme which is regularly immersed in Y. If X satisfies R_1, then X is normal.

Proof. The assertion follows from (III,4.5) and (2.13).

Definition (2.15). - A domain A is said to be factorial (or a unique factorization domain) if every element f has the form Πf_i where the f_i are irreducible elements and the (prime) ideals $f_i A$ are uniquely determined by f. A locally noetherian scheme is said to be locally factorial if the local ring of each point is factorial.

Proposition (2.16). - Let A be a noetherian domain. Then A is factorial if and only if every height 1 prime is principal.

Proof. Suppose A is factorial and let p be a prime of A. If $f = \Pi f_i \in p$ where the f_i are irreducible elements, then $f_i \in p$ for some i Thus, if p has height 1, it follows that $p = f_i A$.

Conversely, let f be a nonzero element of A and $\{f_i A\}$ the set of essential primes of fA having height 1. Choose integers r_i inductively as follows: Given $r_1,\ldots,r_{i-1}$, let r_i be the largest integer such that $\prod_{j=1}^{i} f_j^{r_j} | f$, Then $u = f/\Pi f_j^{r_j} \in A$ and uA is easily seen to have no essential primes of height 1. By Krull's theorem (III,1.10), u is a unit and $f = u^{-1}\Pi f_j^{r_j}$; so, A is factorial.

Remark (2.17). - It is easily seen that a factorial domain is normal.

3. Divisors

Definition (3.1). - Let X be a locally noetherian scheme and $J(X)$ the set of reduced irreducible closed subschemes W of X of codimension 1. A <u>divisorial cycle</u> (Weil divisor) is a formal sum $\sum_{W \in J(X)} n_W W$ in which the set of generic points of those W such that $n_W \neq 0$ is locally finite. An element of $J(X)$ is called a <u>prime divisorial cycle</u>; a divisorial cycle is said to be <u>positive</u> if all $n_W \geq 0$; the group of divisorial cycles is denoted $\mathfrak{Z}^1(X)$.

<u>Definition (3.2)</u>. - Let X be a ringed space. The sheaf of <u>meromorphic functions</u> K_X is defined as the sheaf associated to the presheaf whose sections over an open set U are the elements of the total quotient ring of $\Gamma(U, O_X)$. A (Cartier) <u>divisor</u> D is defined as a global section of the sheaf K_X^*/O_X^*, (where, if A_X is a sheaf of rings, A_X^* denotes the (abelian) sheaf whose sections are the units of A_X). The group of divisors is denoted $\mathrm{Div}(X)$. For each $f \in \Gamma(X, K_X^*)$, let (f) denote the image of f in $\mathrm{Div}(X)$.

<u>Remark (3.3)</u>. - Let X be a ringed space. A divisor D is represented by an open covering $\{U_\alpha\}$ of X and local equations $f_\alpha \in \Gamma(U_\alpha, K_X^*)$ such that $f_\alpha/f_\beta \in \Gamma(U_\alpha \cap U_\beta, O_X^*)$; two such collections $\{U_\alpha, f_\alpha\}$ and $\{V_\beta, g_\beta\}$ represent the same divisor if and only if there exists a common refinement $\{W_\gamma\}$ and elements $h_\gamma \in \Gamma(W_\gamma, O_X^*)$ such that, if $W_\gamma \subset U_\alpha \cap V_\beta$, then $f_\alpha = g_\beta h_\gamma$ on W_γ.

<u>Remark (3.4)</u>.- Let X be a ringed space. A divisor D defines an invertible sheaf $O_X(D)$, contained in K_X: If $\{U_\alpha, f_\alpha\}$ represents D, then $O_X(D)|U_\alpha = f_\alpha^{-1} O_X|U_\alpha \subset K_X|U_\alpha$.

<u>Definition (3.5)</u>. - Let X be a ringed space. A divisor D is said to be <u>effective</u> (positive) if any one of the following equivalent conditions holds:

(i) If $\{U_\alpha, f_\alpha\}$ represents D, then the local equations f_α are sections of O_X.

(ii) $O_X \subset O_X(D) \subset K_X$.

(iii) $O_X(-D)$ is a sheaf of ideals.

Remark (3.6). - Let X be a scheme and D an effective divisor. Then there is an exact sequence

$$0 \longrightarrow O_X(-D) \longrightarrow O_X \longrightarrow O_D \longrightarrow 0$$

and O_D is the structure sheaf of a closed subscheme, denoted Supp(D), (or, simply D).

Definition (3.7). - Let X be a ringed space. The Picard group of X, denoted $\mathrm{Pic}(X)$, is defined as the group of isomorphism classes of invertible sheaves on X.

Remark (3.8). - Let X be a ringed space. It is easily seen that $\mathrm{Pic}(X) = \check{H}^1(X,O_X^*)$ [7] 0_I, 5.4.7). Furthermore, the exact sequence

$$0 \longrightarrow O_X^* \longrightarrow K_X^* \longrightarrow K_X^*/O_X^* \longrightarrow 0$$

yields an exact sequence

$$\Gamma(X,K_X^*) \longrightarrow \mathrm{Div}(X) \xrightarrow{\delta} \mathrm{Pic}(X) \longrightarrow \check{H}^1(X,K_X^*)$$

where $\delta(D) = O_X(D)$. Hence if $\check{H}^1(X,K_X^*) = 0$, then every invertible sheaf comes from a divisor.

Suppose X is noetherian and satisfies S_1. Let A be an affine coordinate ring of X. Then, by (2.2), all essential primes p of A are minimal; so, by (II,4.7), the total quotient ring K of A is artinian and, by (II,4.9), $K = \Pi K_{x_0}$ as x_0 runs through all generic points of Spec(A). Thus, $K_X = \Pi(i_{x_0 *})K'_{x_0}$ where, if x_0 is a generic point of X, then K'_{x_0} is the constant sheaf of K_{x_0} on $\{\bar{x}_0\}$ and $i_{x_0} : \mathrm{Spec}(O_{x_0}) \longrightarrow X$ is the canonical immersion. Therefore, $\Gamma(X,K_X) = \Pi K_{x_0}$ as x_0 runs through the generic points of X and $H^1(X,K_X^*) = 0$.

<u>Definition (3.9)</u>. - Let X be an R_1 locally noetherian scheme. Then the <u>cycle map</u>, $cyc : Div(X) \longrightarrow \mathfrak{Z}^1(X)$, is a homomorphism defined as follows: If W is a prime divisorial cycle, then, at the generic point w of W, the local ring O_w is a discrete valuation ring by (2.6); let v_W be the associated valuation. If $D \in Div(X)$, let $f_w \in K_w^*$ be a local equation of D at w and define $v_W(D)$ as $v_W(f_w)$, and $cyc(D)$ as $\Sigma v_W(D)W$. A divisorial cycle is called <u>locally principal</u> if it is of the form $cyc(D)$.

<u>Proposition (3.10)</u>. - Let X be a normal, locally noetherian scheme and D a divisor. Then:

(i) D is effective if (and only if) $cyc(D)$ is positive.

(ii) cyc is injective.

(iii) cyc is bijective if and only if X is locally factorial.

<u>Proof</u>. Let x be a point of X and $f \in K_x^*$ a local equation of D at x. If $cyc(D) \geqslant 0$, then, for each height 1 prime p of $A = O_x$, $f \in A_p$. So, by (2.12), $f \in A = \cap A_p$ and D is effective. If $cyc(D) = 0$, then both D and $-D$ are effective; hence, $f \in A^*$ and $D = 0$. Thus (i) and (ii) hold.

To prove (iii), let x be a point of X and p a height 1 prime of O_x. Then p defines a prime divisorial cycle W. If $W = cyc(D)$ for some divisor D, let f be a local equation of D at x. Then, by (i) $f \in O_x$. Let $\{q_i\}_{i=1}^r$ be an irredundant primary decomposition of fA (II,3.14). Since A is normal, each essential prime of fA has height 1 by (2.12). By localization (II,3.17), it follows that $r = 1$ and $(f) = p$. Hence, by (2.16), X is locally factorial.

Conversely, suppose X is locally factorial. Then, by (2.16), a prime divisorial cycle W is "cut out" at each $x \in X$ by some element $f_x \in O_x$. The f_x are easily seen to define a divisor D such that $\mathrm{cyc}(D) = W$. By linearity, cyc is therefore surjective.

Lemma (3.11). - Let A be a noetherian local domain of depth ≥ 2. Let $X = \mathrm{Spec}(A)$, x be the closed point of X and $U = X - \{x\}$. If U is locally factorial and $\mathrm{Pic}(U) = 0$, then A is factorial.

Proof. Since U is locally factorial, it is normal; so, by Serre's criterion (2.13), it satisfies R_1 and S_2; hence, since $\mathrm{depth}(A) \geq 2$, X satisfies R_1 and S_2. By (2.13), A is normal.

Any height 1 prime p of A defines a prime divisorial cycle W on X. Since U is locally factorial, $W|U$ is locally principal by (3.10). So, since $\mathrm{Pic}(U) = 0$ and U is reduced, $W|U$ is the divisor of a rational function f by (3.8). By (III,3.15), $\dim(A) \geq \mathrm{depth}(A) \geq 2$. So, since f has no poles on U, f has no poles on X; hence, since A is normal, $f \in A$ by (2.12). Let $\{q_i\}_{i=1}^{r}$ be an irredundant primary decomposition of fA. Since A is normal, each essential prime of fA has height 1 by (2.12). By localization, it follows that $r = 1$ and $fA = p$. Hence, by (2.16), A is factorial.

Proposition (3.12). - Let X be a local ringed space and

$$0 \longrightarrow F' \longrightarrow F \longrightarrow F'' \longrightarrow 0$$

an exact sequence of locally free O_X-Modules of finite rank. Then there exists a canonical isomorphism

$$\Lambda^{\max}F' \otimes \Lambda^{\max}F'' \xrightarrow{\sim} \Lambda^{\max}F.$$

Proof. Choose an open cover $\{U_\alpha\}$ of X such that $F|U_\alpha = F'|U_\alpha \oplus G_\alpha$ where G_α is a free O_{U_α}-Module. The canonical isomorphisms $v_\alpha : G_\alpha \longrightarrow F''|U_\alpha$ and $(\Lambda^{max}F'|U_\alpha)\otimes(\Lambda^{max}G_\alpha) \xrightarrow{\sim} \Lambda^{max}F|U_\alpha$ yield an isomorphism

$$u_\alpha : (\Lambda^{max}F')\otimes(\Lambda^{max}F'')|U_\alpha \longrightarrow \Lambda^{max}F|U_\alpha.$$

It remains to show that u_α and u_β coincide on $U_\alpha \cap U_\beta$.

On $U_\alpha \cap U_\beta$, we have $v_\alpha = v_\beta \circ w_{\beta\alpha}$ where $w_{\beta\alpha} : G_\alpha \longrightarrow G_\beta$ is the "projection parallel to F' " defined as follows: For each section $s \in \Gamma(U_\alpha \cap U_\beta, G_\alpha)$, $w_{\beta\alpha}(s) = s + t_{\beta\alpha}(s)$ with $t_{\beta\alpha}(s) \in \Gamma(U_\alpha \cap U_\beta, F')$. However, then $u_\alpha = u_\beta \circ \det(z_{\beta\alpha})$ where $z_{\beta\alpha} : F' \oplus M_\alpha \longrightarrow F' \oplus M_\beta$ is given by $\begin{pmatrix} id & t_{\beta\alpha} \\ 0 & id \end{pmatrix}$. Thus, $\det(z_{\beta\alpha}) = id$ and $u_\alpha = u_\beta$ on $U_\alpha \cap U_\beta$.

__Lemma (3.13) [7], IV,1.7.7)__. - Let X be a quasi-compact, quasi-separated scheme and U a quasi-compact open subset. Then, for each quasi-coherent $(O_X|U)$-Module F of finite type, there exists a quasi-coherent O_X-Module G of finite type such that $G|U = F$.

__Theorem (3.14) (Auslander-Buchsbaum)__. - A regular local ring A is factorial.

__Proof__. (Kaplansky). If the dimension r of A is zero, then A is a field; if $r = 1$, then, by (2.6), A is principal, so factorial. Assume $r \geqslant 2$. Let $X = \mathrm{Spec}(A)$, x be the closed point of X and $U = X - \{x\}$. If $y \in U$, then O_y is regular by (III,5.15 and 5.16) and $\dim(O_y) < r$; hence, U may be assumed locally factorial by induction on r. Since A is regular, by (III,4.12), $\mathrm{depth}(A) = \dim(A) \geqslant 2$.

Let L be an invertible O_U-Module. By (3.13), there exists a coherent O_X-Module F such that $F|U = L$. Since A is regular, $\mathrm{gl.hd}(A) = r$ by (III,5.11); hence, there exists a resolution

$$0 \longrightarrow O_X^{h_r} \longrightarrow \cdots \longrightarrow O_X^{h_0} \longrightarrow F \longrightarrow 0.$$

It therefore follows from (3.12) that $L = O_U$ Hence, Pic(U) = 0; so, by (3.11), A is factorial.

Corollary (3.15). - Let X be a regular scheme and Y a closed subscheme of pure codimension 1. Then Y is normal if (and only if) Y satisfies R_1.

Proof. The assertion follows immediately from (2.13), (III,4.5) and (III,4.12).

4. Stability

Lemma (4.1). - Let $\varphi : A \longrightarrow B$ be a local homomorphism of noetherian rings, k the residue field of A, and $u : M \longrightarrow N$ a B-homomorphism of finite B-modules. Suppose N is a flat A-module. Then the following conditions are equivalent:

(i) u is injective and C = coker(u) is A-flat.

(ii) $u\otimes 1 : M\otimes_A k \longrightarrow N\otimes_A k$ is injective.

Proof. Assume (i). Then the sequence $0 \longrightarrow M \xrightarrow{u} N \longrightarrow C \longrightarrow 0$ is exact and yields the exact sequence

$$\mathrm{Tor}_1^A(C,k) \longrightarrow M\otimes_A k \xrightarrow{u\otimes 1} N\otimes_A k.$$

Since C is A-flat, $u\otimes 1$ is injective.

Conversely, the exact sequence $0 \longrightarrow u(M) \longrightarrow N \longrightarrow C \longrightarrow 0$ yields the exact sequence

$$0 \longrightarrow \mathrm{Tor}_1^A(C,k) \longrightarrow u(M)\otimes_A k \longrightarrow N\otimes_A k.$$

Assume (ii). Then the natural surjection $M\otimes_A k \longrightarrow u(M)\otimes_A k$ is bijective; so, by the exact sequence, $\mathrm{Tor}_1^A(C,k) = 0$. Hence, by the local criterion (V,3.2), C is flat over A.

Since N and C are flat, it follows that $u(M)$ is flat. Let $K = \ker(u)$. Then the exact sequence, $0 \longrightarrow K \longrightarrow M \longrightarrow u(M) \longrightarrow 0$ yields the exact sequence

$$0 \longrightarrow K\otimes_A k \longrightarrow M\otimes_A k \xrightarrow{u\otimes 1} u(M)\otimes_A k.$$

Since $u\otimes 1$ is injective, $K\otimes_A k = 0$. Since φ is a local homomorphism and $\varphi(m)K = K$, it follows from Nakayama's lemma that $K = 0$.

Proposition (4.2). - Let A, B be noetherian local rings, k the residue field of $A, \varphi : A \longrightarrow B$ a local homomorphism. M a finite A-module and N a finite B-module. Suppose N is a flat A-module. Then

$$\operatorname{depth}_B(M\otimes_A N) = \operatorname{depth}_A(M) + \operatorname{depth}_{B\otimes_A k}(N\otimes_A k)$$

Proof. By (III,3.15), we may assume $M \neq 0$ and $N \neq 0$. Suppose $\operatorname{depth}_A(M) = 0$ and $\operatorname{depth}_{B\otimes_A k}(N\otimes_A k) = 0$. Let m (resp. n) be the maximal ideal of A (resp. B). By (III,3.11), $m \in \operatorname{Ass}_A(M)$ and, by (III,3.11 and 3.16), $n \in \operatorname{Ass}_B(N\otimes_A k)$. By (II,3.2), there exists an exact sequence $0 \longrightarrow k \longrightarrow M$; so, since N is A-flat, the sequence $0 \longrightarrow N\otimes_A k \longrightarrow N\otimes_A M$ is exact. Hence, $n \in \operatorname{Ass}_B(N\otimes_A k) \subset \operatorname{Ass}_B(M\otimes_A N)$ and $\operatorname{depth}(M\otimes_A N) = 0$.

Suppose $\operatorname{depth}_A(M) > 0$. Let $x \in m$ be M-regular, $M' = M/xM$, $N' = N/xN$, $A' = A/xA$ and $B' = B/xB$. Since $N' = N\otimes_A A'$, N' is A'-flat; furthermore, $N'\otimes_{A'} k = N\otimes_A k$ and $M'\otimes_{A'} N' = (M\otimes_A N)/x(M\otimes_A N)$. By (III,3.10 and 3.16), $\operatorname{depth}_{A'}(M') = \operatorname{depth}_A(M)-1$ and $\operatorname{depth}_{B'}(M'\otimes_{A'} N') = \operatorname{depth}_B(M\otimes_A N)-1$. Thus, the formula follows by induction.

Suppose $\operatorname{depth}_{B\otimes_A k}(N\otimes_A k) > 0$. Let $y \in n$ be $(N\otimes_A k)$-regular and $N' = N/yN$ Then (4.1) implies that the sequence

$$0 \longrightarrow N \xrightarrow{y} N \longrightarrow N' \longrightarrow 0$$

is exact and that N' is A-flat; it follows that y is $(M\otimes_A N)$-regular. Since $(N\otimes_A k)/y(N\otimes_A k) \cong N'\otimes_A k$ and $(M\otimes_A N)/y(M\otimes_A N) \cong M\otimes_A N'$, (III,3.10) implies that $\text{depth}_{B\otimes_A k}(N'\otimes_A k) = \text{depth}_{B\otimes_A k}(N\otimes_A k)-1$ and $\text{depth}_B(M\otimes_A N') = \text{depth}_B(M\otimes_A N)-1$. Thus the formula follows by induction.

<u>Proposition (4.3)</u>. - Let $\varphi : A \longrightarrow B$ be a local homomorphism of noetherian rings. Suppose B is flat over A. Then $\text{gl.hd}(A) \leq \text{gl.hd}(B)$.

<u>Proof</u>. We may assume $q = \text{gl.hd}(B)$ is finite. Let M, N be two A-modules. Clearly, $\text{Tor}^A_{q+1}(M,N)\otimes_A B = \text{Tor}^B_{q+1}(M\otimes_A B, N\otimes_A B)$, which is zero by hypothesis. By (V,1.6), B is faithfully flat over A; so, by (V,1.4), $\text{Tor}^A_{q+1}(M,N) = 0$. Hence, by (III,5.7 and 5.9), $\text{gl.hd}(A) \leq q$.

<u>Lemma (4.4)</u>. - Let A be a ring, $A[T]$ the polynomial ring in one variable and M an $A[T]$-module. Then $\text{proj.dim}_{A[T]}(M) \leq \text{proj.dim}_A(M) + 1$.

<u>Proof</u>. Set $M[T] = M\otimes_A A[T]$ and consider the sequence

$$0 \longrightarrow M[T] \xrightarrow{f} M[T] \xrightarrow{g} M \longrightarrow 0$$

where $f(x\otimes a) = x\otimes Ta - Tx\otimes a$ and $g(x\otimes a) = ax$. Clearly, g is surjective and $g \circ f = 0$. If $g(\Sigma x_i\otimes T^i) = 0$, then $\Sigma x_i\otimes T^i = f(\Sigma x_i\otimes T^{i-1} + Tx_i\otimes T^{i-2} + \ldots + T^{i-1}x_i\otimes 1)$; so, the sequence is exact in the middle. If $f(\Sigma x_i\otimes T^i) = 0$, then $x_d\otimes T^{d+1} = 0$ where d is the largest integer such that $x_d\otimes T^d \neq 0$; hence, f is injective and the sequence is exact.

It follows from (III,5.2) that $\text{proj.dim}_{A[T]}(M) \leq \text{proj.dim}_{A[T]}(M[T]) + 1$. Finally, since $A[T]$ is flat, it follows easily from the definition that $\text{proj.dim}_{A[T]}(M[T]) \leq \text{proj.dim}_A(M)$.

Theorem (4.5). - Let A be a regular ring. Then the polynomial ring $A[T_1,\ldots,T_r]$ is regular.

Proof. By induction, we may assume $r = 1$; by (4.4), $\text{gl.hd}(A[T]) \leq \text{gl.hd}(A) + 1$, so the assertion follows from (III,5.18).

Proposition (4.6). - Let $\varphi : A \longrightarrow B$ be a local homomorphism of noetherian rings and M a finite B-module. Let m be the maximal ideal of A, $(x_1,\ldots,x_r)$ an A-regular sequence of m and $I = x_1A + \ldots + x_rA$. Then M is A-flat if (and only if) M/IM is (A/I)-flat and the sequence $(x_1,\ldots, x_r)$ is M-regular.

Proof. By (III,3 4), the homomorphisms $(M/IM)[T_1,\ldots T_r] \longrightarrow gr^*_I(M)$ and $(A/I)[T_1,\ldots,T_r] \longrightarrow gr^*_I(A)$ are bijective; hence, the canonical homomorphism $(M/IM)\otimes_{A/I} gr^*_I(A) \rightarrow gr^*_I(M)$ is bijective Therefore, M is A-flat by the local criterion (V,3.2).

Theorem (4.7). - Let A,B be noetherian local rings, k the residue field of A, and $\varphi : A \longrightarrow B$ a local homomorphism. Then the following conditions are equivalent:

(i) A and B are regular and, if $x_1,\ldots,x_r$ are regular parameters of A, then $y_1 = \varphi(x_1),\ldots,y_r = \varphi(x_r)$ are regular parameters of B.

(ii) B and $B\otimes_A k$ are regular and B is flat over A.

(iii) A and $B\otimes_A k$ are regular and B is flat over A.

(iv) A and $B\otimes_A k$ are regular and $\dim(B) = \dim(A) + \dim(B\otimes_A k)$.

__Proof__. If $r = \dim(A)$, then, by (III,4.11) and (4.6), condition (iii) is equivalent to the condition

(iii') A is regular, and if $x_1,\dots,x_r$ are regular parameters of A, then $y_1 = \varphi(x_1),\dots,y_r = \varphi(x_r)$ form a B-regular sequence and $B/(y_1B + \dots + y_rB)$ is regular.

Now (i) and (iv) are equivalent by (III,4.10); furthermore, (iii) implies (iv) by (V,2.11) and (i) implies (iii') by (III,4.11 and 4.10). Hence, (i), (iii) and (iv) are equivalent. Clearly, (i) and (iii) together imply (ii) and (ii) implies (iii) by (4.3) and (III,5.11 and 5.15).

__Theorem (4.8)__. - Let X,Y be locally noetherian schemes and $f : X \longrightarrow Y$ a faithfully flat morphism. Then:

(i) If X satisfies R_k (resp. S_k), then Y satisfies R_k (resp. S_k).

(ii) Suppose that, for each $y \in f(X)$, the scheme $f^{-1}(y)$ satisfies R_k (resp. S_k). If Y satisfies R_k (resp. S_k), then X satisfies R_k (resp. S_k).

__Proof__. To prove (i), let y be a point of Y and x a generic point of $f^{-1}(y)$. Then, $\dim(O_x \otimes_{O_y} k(y)) = 0$; so, by (V,2.11), $\dim(O_x) = \dim(O_y)$. However, if O_x is regular, then, by (4.3), O_y is regular; thus, if X satisfies R_k, then Y satisfies R_k. Furthermore, by (III,3.15), $\operatorname{depth}(O_x \otimes_{O_y} k(y)) = 0$; so, by (4.2), $\operatorname{depth}(O_x) = \operatorname{depth}(O_y)$; thus, if X satisfies S_k, then Y satisfies S_k.

To prove (ii), let x be a point of X and $y = f(x)$. Then it suffices to show that, if $\dim(O_x) \leq k$, then O_x is regular (resp. that $\operatorname{depth}(O_x) \geq \inf\{k,\dim(O_x)\}$). Since f is flat, by (V,2.11), $\dim(O_x) = \dim(O_y) + \dim(O_x \otimes_{O_y} k(y))$ (resp. by (4.2), $\operatorname{depth}(O_x) = \operatorname{depth}(O_y) + \operatorname{depth}(O_x \otimes_{O_y} k(y))$); hence, if $\dim(O_x) \leq k$, then, *a fortiori*, $\dim(O_y) \leq k$ and $\dim(O_x \otimes_{O_y} k(y)) \leq k$, and, by hypothesis, O_y and $O_x \otimes_{O_y} k(y)$ are regular. So, by (4.7), O_x is regular; thus, X satisfies R_k. Similarly, $\mathrm{depth}(O_x) \geq \inf\{k, \dim(O_y)\} + \inf\{k, \dim(O_x \otimes_{O_y} k(y))\} \geq \inf\{k, \dim(O_x)\}$; thus, X satisfies S_k.

Theorem (4.9). - Let X,Y be locally noetherian schemes and $f : X \longrightarrow Y$ a surjective, smooth morphism. Then X satisfies R_k (resp. S_k) if and only if Y satisfies R_k (resp. S_k). Consequently, X is generically reduced (resp. without embedded components, reduced, regular, Cohen-Macaulay, normal) if and only if Y is.

Proof. Since f is faithfully flat, the assertion follows easily from (4.8), (4.5), (III,4.12), (2.2) and (2.13).

5. Differential properties

Theorem (5.1). - Let S be a locally noetherian scheme, X,Y two schemes locally of finite type over S and $f : X \longrightarrow Y$ an S-morphism. Suppose f is smooth at $x \in X$. Then:

(i) At x, the sequence $0 \longrightarrow f^*\Omega^1_{Y/S} \longrightarrow \Omega^1_{X/S} \longrightarrow \Omega^1_{X/Y} \longrightarrow 0$ is exact and split.

(ii) At x, $\Omega^1_{X/Y}$ is free of rank $n = \dim_x(f)$.

Proof. Since all properties are local on X, we may assume f is a composition $X \xrightarrow{g} \mathbb{A}^n_Y \xrightarrow{p} Y$ where g is étale. By (VI,1.19), the sequence $0 \longrightarrow p^*\Omega^1_{Y/S} \longrightarrow \Omega^1_{\mathbb{A}^n_Y/S} \longrightarrow \Omega^1_{\mathbb{A}^n_Y/Y} \longrightarrow 0$ exact and split.

Applying g^*, we obtain the split, exact sequence

$$0 \longrightarrow g^*p^*\Omega^1_{Y/S} \longrightarrow g^*\Omega^1_{\mathbb{A}^n_Y/S} \longrightarrow g^*\Omega^1_{\mathbb{A}^n_Y/Y} \longrightarrow 0.$$

However, $g^*p^*\Omega^1_{Y/S} = f^*\Omega^1_{Y/S}$, and, since g is étale, $g^*\Omega^1_{\mathbb{A}^n_Y/S} \xrightarrow{\sim} \Omega^1_{X/S}$

and $g^*\Omega^1_{A^n_Y/Y} \xrightarrow{\sim} \Omega^1_{X/Y}$ by (VI,4.9); whence (i). Finally, it follows from (VI,1.4) that $\Omega^1_{X/Y}$ is free of rank n.

Proposition (5.2). - Let S be a locally noetherian scheme, X,Y two schemes locally of finite type over S and $g : X \longrightarrow Y$ an S-morphism. Suppose X and Y are smooth over S. Then g is étale at $x \in X$ if (and only if) the canonical map $g^*\Omega^1_{Y/S} \longrightarrow \Omega^1_{X/S}$ is an isomorphism at x.

Proof. The conditions are local, so we may assume that X and Y are affine and that the map $g^*\Omega^1_{Y/S} \longrightarrow \Omega^1_{X/S}$ is an isomorphism. By (VI,1.6), $\Omega^1_{X/Y} = 0$; hence, by (VI,3.3), g is unramified at x. Thus, it remains to prove g is flat. Let s be the image of x in S and $k = k(s)$. By (VI,4.8), we may assume $S = \mathrm{Spec}(k)$ and that X and Y are algebraic k-schemes. By (V,5.5), g is flat on an open set; hence, the closed points of an algebraic scheme being dense (III,2.8), we may assume x is closed. Since k is regular, X and Y are regular by (4.8). Since g is quasi-finite, it suffices, by (V,3.6) to show that $\dim(O_x) = \dim(O_{g(x)})$. Since x is closed, it follows from (III,2.6) that $\dim_x(X/S) = \dim(O_x)$ and $\dim_{g(x)}(Y/S) = \dim(O_{g(x)})$. The contention now follows from (5.1,(ii)) and the hypothesis.

Theorem (5.3). - Let S be a locally noetherian scheme, X,Y two schemes locally of finite type over S and $f : X \longrightarrow Y$ an S-morphism locally of finite type. Let x be a point of X and $y = f(x)$. Suppose Y is smooth over S at y. Then f is smooth of x if and only if the following conditions hold:

(a) At x, X is smooth over S.

(b) At x, the sequence $0 \longrightarrow f^*\Omega^1_{Y/S} \longrightarrow \Omega^1_{X/S} \longrightarrow \Omega^1_{X/Y} \longrightarrow 0$ is exact.

(c) At x, $\Omega^1_{X/Y}$ is free of rank $n = \dim_x(f)$.

Proof. The necessity follows from (1.7) and (5.1). Conversely, take $g_{1,x},\ldots,g_{n,x} \in O_x$ such that $dg_{1,x},\ldots,dg_{n,x}$ form a basis of $(\Omega^1_{X/Y})_x$. Since the conditions are local, we may assume that the $g_{i,x}$ extend to global sections g_i of X. The g_i define a morphism g such that the following diagram commutes.

$$\begin{array}{ccc} X & \xrightarrow{\ g\ } & \mathbb{A}^n_Y \\ & \searrow^{f} & \downarrow p \\ & & Y \end{array}$$

It remains to show that g is étale. Consider the exact sequence $0 \longrightarrow p^*\Omega^1_{Y/S} \longrightarrow \Omega^1_{\mathbb{A}^n_Y/S} \longrightarrow \Omega^1_{\mathbb{A}^n_Y/Y} \longrightarrow 0$; applying g^*, we obtain the diagram

$$\begin{array}{ccccccccc} 0 & \longrightarrow & g^*p^*\Omega^1_{Y/S} & \longrightarrow & g^*\Omega^1_{\mathbb{A}^n_Y/S} & \longrightarrow & g^*\Omega^1_{\mathbb{A}^n_Y/Y} & \longrightarrow & 0 \\ & & \downarrow \wr & & \downarrow \alpha & & \downarrow \beta & & \\ 0 & \longrightarrow & f^*\Omega^1_{Y/S} & \longrightarrow & \Omega^1_{X/S} & \longrightarrow & \Omega^1_{X/Y} & \longrightarrow & 0 \end{array}$$

By construction, β is an isomorphism; hence, by the five lemma, α is an isomorphism and g is étale by (5.2).

Definition (5.4). - Let $f : X \longrightarrow Y$ be a morphism of schemes. The tangent space of X/Y at $x \in X$, denoted $T_{X/Y}(x)$, is defined as the $k(x)$-vector space $\mathrm{Hom}_{k(x)}(\Omega^1_{X/Y}(x), k(x))$, (where $\Omega^1_{X/Y}(x) = \Omega^1_{X/Y} \otimes_{O_X} k(x)$).

Corollary (5.5). - Let S be a locally noetherian scheme, X,Y schemes locally of finite type over S and $f : X \longrightarrow Y$ an S-morphism. Let x be a point of X and $y = f(x)$. Suppose X (resp. Y) is smooth over S at x (resp. y).

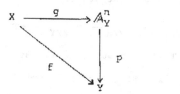

Then f is smooth at x if and only if $T_x(f)$: $T_{X/S}(x) \longrightarrow T_{Y/S}(y) \otimes_{k(y)} k(x)$ is surjective. In particular, if x is rational over $k(y)$, f is smooth at x if and only if $df(x) : T_{X/S}(x) \longrightarrow T_{Y/S}(y)$ is surjective.

Proof. By (VI,1.6), the sequence

$$f^*\Omega^1_{Y/S} \longrightarrow \Omega^1_{X/S} \longrightarrow \Omega^1_{X/Y} \longrightarrow 0$$

is exact. Assume $T_x(f)$ is surjective. By (5.1), $\Omega^1_{X/S}$ and $f^*\Omega^1_{Y/S}$ are free at x. So it follows from (IV,3.2) and Nakayama's lemma that the sequence

$$0 \longrightarrow \underline{\mathrm{Hom}}_{O_X}(\Omega^1_{X/Y}, O_X) \longrightarrow \underline{\mathrm{Hom}}_{O_X}(\Omega^1_{X/S}, O_X) \longrightarrow \underline{\mathrm{Hom}}_{O_X}(f^*\Omega^1_{Y/S}, O_X) \longrightarrow 0$$

is exact at x. It follows that, at x, the sequence splits and $\underline{\mathrm{Hom}}_{O_X}(\Omega^1_{X/Y}, O_X)$ is free; hence, we have the commutative diagram with exact rows

$$\begin{array}{ccccccc}
f^*\Omega^1_{Y/S} & \longrightarrow & \Omega^1_{X/S} & \longrightarrow & \Omega^1_{X/Y} & \longrightarrow & 0 \\
\downarrow \alpha & & \downarrow \beta & & \downarrow \gamma & & \\
0 \longrightarrow (f^*\Omega^1_{Y/S})^{\vee\vee} & \longrightarrow & (\Omega^1_{X/S})^{\vee\vee} & \longrightarrow & (\Omega^1_{Y/X})^{\vee\vee} & \longrightarrow & 0
\end{array}$$

where $F^\vee = \underline{\mathrm{Hom}}_{O_X}(F, O_X)$ for any locally free O_X-Module of finite rank.

Then, at x, α and β are isomorphisms, so γ is an isomorphism by the five lemma; hence, $\Omega^1_{X/Y}$ is free and $f^*\Omega^1_{Y/S} \rightarrow \Omega^1_{X/S}$ is injective. Hence, by (5.3), f is smooth at x. The converse is similar.

Lemma (5.6). - Let S be a locally noetherian scheme, X an S-scheme locally of finite type, x a point of X and $g_1, \ldots, g_n$ global sections of O_X. Suppose X is smooth over S at x. Then the following conditions are equivalent:

(i) $g_1,\dots,g_n$ define an S-morphism $g\colon X \longrightarrow \mathbb{A}^n_S$ which is étale at x.

(ii) $dg_1,\dots,dg_n$ form a basis of $\Omega^1_{X/S}$ at x.

(iii) $dg_1(x),\dots,dg_n(x)$ form a basis of $\Omega^1_{X/S}(x)$.

Proof. Note that the map $g^*\Omega^1_{\mathbb{A}^n_{S/S}} \longrightarrow \Omega^1_{X/S}$ is an isomorphism at x if and only if (ii) (or, equivalently, (iii)) holds and apply (5.2).

Theorem (5.7). - Let S be a locally noetherian scheme, X an S-scheme locally of finite type, Y a closed S-subscheme, and J its sheaf of ideals. Let x be a point of Y and $g_1,\dots,g_n$ global sections of O_X. Suppose X is smooth over S at x. Then the following conditions are equivalent:

(i) There exists an open neighborhood X_1 of x such that $g_1,\dots,g_n$ define an étale morphism $g : X_1 \longrightarrow \mathbb{A}^n_S$ and $g_1,\dots,g_p$ generate J on X_1; i.e., $Y_1 = Y \cap X_1$ is the fiber over a linear subscheme $\mathbb{A}^{n-p}_S$ of $\mathbb{A}^n_S$.

(ii) (a) Y is smooth over S at x.

(b) $g_{1,x},\dots,g_{p,x} \in J_x$.

(c) $dg_1(x),\dots,dg_n(x)$ form a basis of $\Omega^1_{X/S}(x)$.

(d) $dg_{p+1}(x),\dots,dg_n(x)$ form a basis of $\Omega^1_{Y/S}(x)$.

(iii) $g_{1,x},\dots,g_{p,x}$ generate J_x and $dg_1(x),\dots,dg_n(x)$ form a basis of $\Omega^1_{X/S}(x)$.

(iv) Y is smooth over S at x, $g_{1,x},\dots,g_{p,x}$ form a minimal set of generators of J_x and $dg_{p+1}(x),\dots,dg_n(x)$ form a basis of $\Omega^1_{Y/S}(x)$.

Furthermore, if these conditions hold, then, at x, the sequence

(5.7.1) $$0 \longrightarrow J/J^2 \longrightarrow \Omega^1_{X/S} \otimes_{O_X} O_Y \longrightarrow \Omega^1_{Y/S} \longrightarrow 0$$

is exact and composed of free O_Y-Modules with bases induced by $\{g_1,\ldots,g_p\}$, $\{dg_1,\ldots,dg_n\}$ and $\{dg_{p+1},\ldots,dg_n\}$.

<u>Proof</u>. Assume (i). Since g is étale, Y_1 is étale over $\mathbb{A}_S^{n-p}$ by (VI,4.7). Thus Y is smooth over S at x with relative dimension n-p. By (5.6), $dg_1,\ldots,dg_n$ form a basis $\Omega^1_{X/S}$ at x and $dg_{p+1},\ldots,dg_n$ form a basis of $\Omega^1_{Y/S}$ at x; so, (ii) and (iii) hold. It follows that $g_1,\ldots,g_p$ are linearly independent elements of J/J^2 at x; since they generate, they are a basis. Therefore, (iv) holds and (5.7.1) is an exact sequence of free O_Y-Modules at x.

Assume (ii) and let X_1 be an open neighborhood of x on which $g_1,\ldots,g_p$ generate J. Consider the commutative diagram

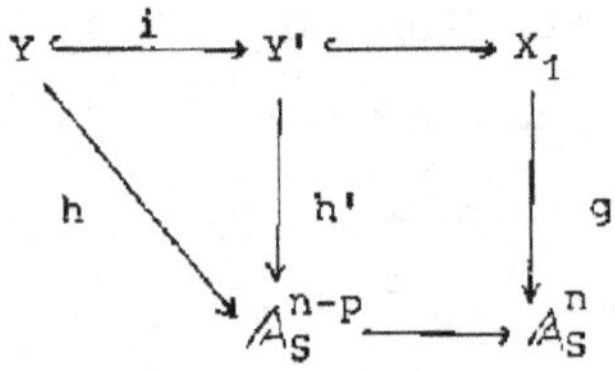

where $Y' = g^{-1}(\mathbb{A}_S^{n-p})$. By (5.6), g and h are étale and, by (VI,3.5), h' is unramified. Hence, by (VI,4.7), i is étale. However, by (VI,5.6), the closed immersion i is open. Therefore $Y = Y'$ and (i) holds.

Assume (iii) and let X_1 be an open neighborhood of x on which $g_1,\ldots,g_p$ generate J and $dg_1,\ldots,dg_n$ form a basis of $\Omega^1_{X/S}$. Then (i) holds by (5.6).

Finally, the implication (iv)$\Longrightarrow$(i) follows from (5.3) and the implication (i)$\Longrightarrow$(ii) of the following theorem.

<u>Theorem (5.8)</u>. - Let S be a locally noetherian scheme, X an S-scheme locally of finite type, Y a closed subscheme of X, J

its sheaf of ideals, x a point of Y and $n = \dim_x(X/S)$. Suppose X is smooth over S at x. Then the following assertions are equivalent:

(i) Y is smooth over S at x and $\dim_x(Y/S) = n-p$.

(ii) There exists an open neighborhood X_1 of x and an étale morphism $g : X_1 \longrightarrow \mathbb{A}_S^n$ such that $X_1 \cap Y = g^{-1}(\mathbb{A}_S^{n-p})$.

(iii) There exist generators $g_{1,x}, \ldots, g_{p,x} \in J_x$ such that $dg_1(x), \ldots, dg_p(x)$ are linearly independent in $\Omega^1_{X/S}(x)$.

(iv) At x, $0 \longrightarrow J/J^2 \longrightarrow \Omega^1_{X/S} \otimes_{O_X} O_Y \longrightarrow \Omega^1_{Y/S} \longrightarrow 0$ is an exact sequence of free O_Y-Modules of ranks p, n, and $n-p$.

Proof. To prove the implication (i)$\Rightarrow$(ii), note that, by (5.1), $\Omega^1_{X/S}$ and $\Omega^1_{Y/S}$ are free at x with ranks n and $n-p$. Take $g_{p+1,x}, \ldots, g_{n,x} \in O_X$ such that $dg_{p+1}(x), \ldots, dg_n(x)$ form a basis of $\Omega^1_{Y/S}(x)$. By (VI,1.8), the sequence

$$J/J^2 \longrightarrow \Omega^1_{X/S} \otimes_{O_X} O_Y \longrightarrow \Omega^1_{Y/S} \longrightarrow 0$$

is exact, so extend $dg_{p+1}(x), \ldots, dg_n(x)$ to a basis $dg_1(x), \ldots, dg_n(x)$ of $\Omega^1_{X/S}(x)$ with $g_{1,x}, \ldots, g_{p,x} \in J_x$. Then it follows from (ii)$\Rightarrow$(i) of (5.7) that (ii) holds.

The implications(ii)$\Rightarrow$(i), (iii), (iv) follow directly from (5.7); the implications(iii) $\Rightarrow$(i), (iv) follow from (5.7) if we extend $dg_1(x), \ldots, dg_p(x)$ to a basis of $\Omega^1_{X/S}(x)$.

Assume (iv) and take $g_{1,x}, \ldots, g_{p,x} \in J_x$ whose residue classes are linearly independent in J_x/J_x^2. By Nakayama's lemma, the $g_{i,x}$ generate J_x, and the exactness of (5.7.1) implies that $dg_1(x), \ldots, dg_p(x)$ are linearly independent. Hence, (iii) holds.

Corollary (5.9). - Let S be a locally noetherian scheme, X an S-scheme locally of finite type, Y a closed subscheme of X, J

its sheaf of ideals, x a point of Y, $n = \dim_x(X/S)$, $g_1,\ldots,g_p$ sections of J over a neighborhood of x. Suppose X and Y are smooth at x. Then the following conditions are equivalent:

(i) $p = \dim_x(X/S) - \dim_x(Y/S)$ and $dg_1(x),\ldots,dg_p(x)$ are linearly independent in $\Omega^1_{X/S}(x)$.

(ii) $g_{1,x},\ldots,g_{p,x}$ generate J_x and $dg_1(x),\ldots,dg_p(x)$ are linearly independent in $\Omega^1_{X/S}(x)$.

(iii) $g_1,\ldots,g_p$ induce a basis of J_x/J_x^2.

(iv) $g_{1,x},\ldots,g_{p,x}$ form a minimal set of generators of J_x.

(v) There exist sections $g_{p+1},\ldots,g_n$ of O_X over some open neighborhood X_1 of x which, together with $g_1,\ldots,g_p$, define an étale morphism $g : X_1 \longrightarrow \mathbb{A}^n_S$ such that $Y \cap X_1 = g^{-1}(\mathbb{A}^{n-p}_S)$.

Proof. Assertions (iii), and (iv) are equivalent by Nakayama's lemma; (i), (ii), (iii) and (v), by (5.7).

Corollary (5.10). - Let S be a locally noetherian scheme, X an S-scheme locally of finite type and Y a hypersurface defined by a global section g of O_X. Assume X is smooth over S at $x \in Y$. Then Y is smooth over S at x if and only if $dg(x) \neq 0$.

Proof. The necessity follows from (iv) $\Longrightarrow$ (ii) of (5.9); the sufficiency, from (iii) $\Longrightarrow$ (i) of (5.8).

Corollary (5.11). - Let S be a locally noetherian scheme and Y an S-scheme locally of finite type over S. Consider a cartesian diagram

$$\begin{array}{ccc} Y' & \longrightarrow & Y \\ \downarrow & \square & \downarrow \\ S' & \longrightarrow & S \end{array}$$

in which $S' \longrightarrow S$ is flat. Let x' be a point of Y' and $x \in Y$,

$s' \in S'$, $s \in S$ its images. Then Y is smooth over S at x if and only if Y' is smooth over S' at x'. In particular, if $S' \longrightarrow S$ is faithfully flat, then Y is smooth over S if and only if Y' is smooth over S'.

Proof. We may assume that S and Y are affine and that $Y \longrightarrow S$ is of finite type. Then there exists a closed immersion $Y \hookrightarrow X = \mathbb{A}^n_S$; let $Y' \hookrightarrow X' = \mathbb{A}^n_{S'}$ be its base extension and let J and J' be the defining sheaves of ideals. Consider the sequences

(5.9.1) $$0 \longrightarrow J/J^2 \longrightarrow \Omega^1_{X/S} \otimes_{O_X} O_Y \longrightarrow \Omega^1_{Y/S} \longrightarrow 0$$

(5.9.2) $$0 \longrightarrow J'/J'^2 \longrightarrow \Omega^1_{X'/S'} \otimes_{O_{X'}} O_{Y'} \longrightarrow \Omega^1_{Y'/S'} \longrightarrow 0$$

Since, by (V,1.6), $O_s \longrightarrow O_{s'}$ is faithfully flat, by (VI,4.10) and (VI,1.18), (5.9.1) is exact if and only if (5.9.2) is exact. Thus, the assertion follows from (iv)$\Longleftrightarrow$(i) of (5.8) and the following lemma.

Lemma (5.12). - Let $\varphi : A \longrightarrow B$ be a local homomorphism of noetherian rings and M a finite A-module. Suppose B is flat over A. Then M is free over A if (and only if) $M \otimes_A B$ is free over B.

Proof. The assertion follows immediately from (V,1.5,(iv)) and (III,5.8).

Theorem (5.13). - Let S be a locally noetherian scheme, X a scheme locally of finite type over S and Y a closed S-subscheme of X. Suppose Y is smooth over S at x. Then X is smooth over S at x if and only if Y is regularly immersed in X at x.

Proof. If X is smooth over S at x, then, by (5.8), there exists an open neighborhood X_1 of x in X and an étale morphism $g : X_1 \longrightarrow \mathbb{A}^n_S$ such that $Y_1 = Y \cap X_1 = g^{-1}(\mathbb{A}^{n-p}_S)$. Since $\mathbb{A}^{n-p}_S$ is regularly immersed in $\mathbb{A}^n_S$ and since g is flat, it follows that Y is regularly immersed in X at x.

Conversely, if Y is regularly immersed in X at x, let $(g_{1,x},\ldots,g_{p,x})$ be an O_x-regular sequence which generates the ideal J_x of Y at x and let $g_{p+1,x},\ldots,g_{n,x}$ be elements of $O_{X,x}$ whose images in $O_{Y,x}$ define an étale morphism $Y \longrightarrow \mathbb{A}^{n-p}_S$. Since the question is local, we may assume the $g_{i,x}$ extend to global sections of X. Then they define a map $g:X \to \mathbb{A}^n_S = X'$, and, in view of (VI,4.6), it remains to show that g is étale at x. The fiber of g at x is identical to the fiber of $g|Y$ at x; thus, g is unramified at x. Applying (4.6) to $A = O_{X',g(x)}$, $M = B = O_{X,x}$ and $I = J_x$, we conclude that g is flat at x.

<u>Theorem (5.14) (Jacobian criterion)</u>. - Let S be an noetherian affine scheme with ring A, Y a closed subscheme of $X = \mathbb{A}^n_S$ and x a point of Y. Let $I = g_1R + \ldots + g_NR$ be the ideal in $A[T_1,\ldots,T_n]=R$ defining Y and $\frac{\partial(g_1,\ldots,g_N)}{\partial(T_1,\ldots,T_n)}(x)$ the matrix whose (i,j)th entry is $\frac{\partial g_i}{\partial T_j}(x)$, (called the <u>Jacobian matrix</u>). The following conditions are equivalent:

(i) Y is smooth over S at x and $\dim_x(Y/S) = n-p$.

(ii) There exists a re-indexing of $g_1,\ldots,g_N$ such that $g_{1,x},\ldots,g_{p,x}$ generate I_x and $\operatorname{rank}\left[\frac{\partial(g_1,\ldots,g_p)}{\partial(T_1,\ldots,T_n)}(x)\right] = p$.

(iii) Y is flat over S at x, $\dim_x(Y/S) = n-p$ and

$$\operatorname{rank}\left[\frac{\partial(g_1,\ldots,g_N)}{\partial(T_1,\ldots,T_n)}(x)\right] = p.$$

Furthermore, if Y is smooth at x and $\dim_x(Y/S) = n-p$. then $g_{1,x},\ldots,g_{p,x} \in I_x$ generate if and only if

$$\operatorname{rank}\left[\frac{\partial(g_1,\ldots,g_p)}{\partial(T_1,\ldots,T_n)}(x)\right] = p.$$

<u>Proof</u>. Assume (i) and, by (5.8), re-index the $g_1,\dots,g_N$ so that $g_1,\dots,g_p$ yields a base of I_x/I_x^2. By (5.9), $g_{1,x},\dots,g_{p,x}$ generate I_x and $dg_1(x),\dots,dg_p(x)$ are linearly independent. Assertion (ii) now results from the following lemma.

<u>Lemma (5.15)</u>. - Let A be a ring, x a point of $\mathbb{A}_A^n$ and $g_1,\dots,g_p \in A[T_1,\dots,T_n]$. Then $dg_1(x),\dots,dg_n(x)$ are linearly independent if and only if $\text{rank}\left[\frac{\partial(g_1,\dots,g_p)}{\partial(T_1,\dots,T_n)}(x)\right] = p$.

<u>Proof</u>. Since $dg_i(x) = \sum \frac{\partial g_i}{\partial T_j}(x)dT_j(x)$ and the $dT_j(x)$ are linearly independent, the assertion follows from the definition rank.

Assume (ii) of (5.14). Then (5.15) implies that $dg_1(x),\dots,dg_p(x)$ are linearly independent; so, by (5.8), it follows that (i) holds.

Trivially, (iii) follows from (i) and (ii) together;it remains to prove that (ii) follows from (iii). By re-indexing $g_1,\dots,g_N$, we may assume $\text{rank}\left[\frac{\partial(g_1,\dots,g_p)}{\partial(T_1,\dots,T_p)}(x)\right] = p$. Let Y' be the subscheme defined by the ideal $g_1R + \dots + g_pR$ By (ii)$\Longrightarrow$(i), Y' is smooth at x. Since Y is flat over S, by (1.9), we may assume $S = \text{Spec}(k(s))$ where s is the image of x in S. Then Y' is reduced by (4.9) and by (5.8) $\dim_x(Y'/S) = n-p$. Since Y is a closed subscheme of Y' and $\dim_x(Y/S) = n-p$, it follows that $Y = Y'$ near x, proving (ii) and necessity in the last assertion. Conversely, in the last assertion, if $g_1,\dots,g_p$ generate, then we may take $N = p$; thus, $\text{rank}\left[\frac{\partial(g_1,\dots,g_p)}{\partial(T_1,\dots,T_n)}(x)\right] = p$ by (i)$\Longrightarrow$(ii).

<u>Proposition (5.16)</u>. - Let S be a locally noetherian scheme, X,Y two S-schemes locally of finite type, $g : X \longrightarrow Y$ an S-morphism,

x a point of X and $y = g(x)$. Assume either of the following two conditions:

(a) $\dim_x(X/S) = \dim_y(Y/S)$, X is flat over S at x and Y is smooth over S at x.

(b) Y is regular at y and $\dim(O_x) = \dim(O_y)$.

Then the following conditions are equivalent:

(i) g is étale at x.

(ii) $g^*\Omega^1_{Y/S} \longrightarrow \Omega^1_{X/S}$ is an isomorphism at x.

(iii) $g^*\Omega^1_{Y/S} \longrightarrow \Omega^1_{Y/S}$ is surjective at x.

Proof. The implication (i)$\Longrightarrow$(ii) was proved in (VI,4.9) and (ii)$\Longrightarrow$(iii) is trivial. Assume (iii). By (VI,1.6) and (VI,3.3), it follows that g is unramified and it remains to prove that g is flat. Under assumption (a), X and Y are flat over S at x; so, by (VI,4.8), we may assume $S = \mathrm{Spec}(k(s))$ where s is the image of x in S. Then, by (4.9), O_y is regular. Since by (V,5.5), g is flat on an open set and since by (III,2.8), the closed points of X are dense, we may assume x (and, therefore y) is closed. Therefore, $\dim(O_x) = \dim_x(X/S)$ and $\dim(O_y) = \dim_y(Y/S)$; so, it suffices to prove that g is flat at x under assumption (b).

By (VI,6.12), O_x is a quotient of a local, étale extension B of O_y. Since O_y is regular of dimension $n = \dim(O_y)$, it follows from (V,2.11) and (4.9) that B is regular of dimension n. Therefore, since $\dim(B) = \dim(O_x)$, it follows that $B = O_x$.

6. Algebraic schemes

Proposition (6.1). - Let k be a field, X an algebraic k-scheme, x a closed point of X, $n = \dim_x(X/k)$ and $g_1,\ldots,g_n$ global sections of O_X. Then the following conditions are equivalent:

(i) $g_1,\ldots,g_n$ define a morphism $g : X \longrightarrow \mathbb{A}^n_k$ which is étale at x.

(ii) $dg_1,\ldots,dg_n$ form a basis of $\Omega^1_{X/k}$ at x.

(iii) $dg_1,\ldots,dg_n$ generate $\Omega^1_{X/k}$ at x.

If, in addition, $k(x)$ is a separable extension of k and $g_{1,x},\ldots,g_{n,x} \in m_x$, then (i), (ii), and (iii) are equivalent to:

(iv) $g_{1,x},\ldots,g_{n,x}$ generate m_x.

<u>Proof</u>. The equivalence of (i), (ii), and (iii) results from (5.16). Under the additional hypotheses, by (VI,3.4), $\Omega^1_{k(x)/k} = 0$; so, the sequence $m_x/m_x^2 \longrightarrow \Omega^1_{O_x/k} \otimes_k k(x) \longrightarrow 0$ is exact by (VI,1.8); thus (iii) follows from (iv).

Conversely, assume (i), (ii) and the additional hypotheses. Then, by definition, Spec(k(x)) and X are smooth over k at x; so by (5.8), the sequence $0 \longrightarrow m_x/m_x^2 \longrightarrow \Omega^1_{O_x/k} \otimes_k k(x) \longrightarrow 0$ is exact; whence, (iv).

<u>Corollary (6.2)</u>. - Let X be an algebraic k-scheme and x a closed point of X. Suppose X is smooth over k at x. Then O_x is regular. Conversely, if k(x) is a separable extension of k and O_x is regular, then X is smooth over k at x.

<u>Proof</u>. The first assertion follows from (4.9). Conversely, applied to a regular system of parameters $g_{1,x},\ldots,g_{n,x} \in m_x$, (6.1) implies the assertion.

<u>Proposition (6.3)</u>. - Let X be an algebraic k-scheme. If X is smooth over k, then X is regular. Conversely, if X is regular and k is perfect, then X is smooth over k.

<u>Proof</u>. The first assertion follows from (4.9). Conversely, if k is perfect and X is regular, the open set U on which X

is smooth contains all closed points by (6.2); hence, by (III,2.8), $U = X$.

Theorem (6.4). - Let k be a field, X an algebraic k-scheme, x a closed point of X and $n = \dim_x(X/k)$. Then the following conditions are equivalent:

(i) X is smooth over k at x.

(ii) $\Omega^1_{X/k}$ is free of rank n at x.

(iii) $\Omega^1_{X/k}$ is generated by n elements at x.

(iv) There exists an open neighborhood U of x such that $U \otimes_k L$ is regular for all field extensions L of k.

(iv') There exists an open neighborhood U of x and a perfect extension k' of k such that $U \otimes_k k'$ is regular.

Proof. The implication (i) $\Longrightarrow$ (ii) follows from (5.1); (iii) $\Longrightarrow$ (i), from (6.1). If X is smooth over k at x, then there exists an open neighborhood U of x on which X is smooth over k; by (1.7), $U \otimes_k L$ is smooth over L and by (4.9), $U \otimes_k L$ is regular. Thus, (i) $\Longrightarrow$ (iv). Finally, the implication (iv') $\Longrightarrow$ (i) follows from (6.3) and (5.11).

Proposition (6.5). - Let k be a field, K an artinian local ring which is a localization of a k-algebra of finite type, m the maximal ideal of K, and $n = \text{tr.deg}_k K/m$. Then the following conditions are equivalent:

(i) K is a finite separable field extension of a purely transcendental extension of k.

(ii) $\Omega^1_{K/k}$ is a free K-module of rank n.

(iii) $\Omega^1_{K/k}$ is a K-module with n generators.

(iv) For all field extensions L of k, $K \otimes_k L$ is reduced.

(iv') There exists a perfect extension k' of k such that $K\otimes_k k'$ is reduced.

Furthermore, K is a finite separable field extension of $k(t_1,\ldots,t_n)$ if and only if $dt_1,\ldots,dt_n$ form a basis of $\Omega^1_{K/k}$.

Proof. Consider K as the local ring of a generic point x of an algebraic k-scheme X. Then, by (6.4), (ii) and (iii) are equivalent and (iii) implies (i) and (iv).

Assume $K = k(t_1,\ldots,t_n,\ldots,t_r)$ is a finite separable extension of $k(t_1,\ldots,t_n)$ and let $X = \mathrm{Spec}(k[t_1,\ldots,t_n,\ldots,t_r])$. Then $t_1,\ldots,t_n$ define a morphism $X \longrightarrow \mathbb{A}^n_k$ which is étale at x (where $O_x = K$); so, by (VI,4.6) and (6.1), (i) implies (ii) and necessity in the last assertion. It remains to prove that (iv') implies (ii) and sufficency in the last assertion.

Assume (iv'). Then, since every element of m is nilpotent by (II,4.7) and since $K\otimes_k k'$ is reduced, K is a field. Let $t_1,\ldots,t_r$ be elements of K such that $dt_1,\ldots,dt_r$ form a basis of $\Omega^1_{K/k}$, and let $L = k(t_1,\ldots,t_r)$. Then $\Omega^1_{L/k}\otimes_L K \xrightarrow{\sim} \Omega^1_{K/k}$; so, by (VI,1.6), $\Omega^1_{K/L} = 0$. Therefore, by (VI,3.3), K is a finite separable extension of L and thus $r \geqslant n$.

Let $f \in k[T_1,\ldots,T_r]$ be a nonzero polynomial of minimal degree such that $f(t_1,\ldots,t_r) = 0$. Then $\sum \frac{\partial f}{\partial T_i}(t)\,dt_i = 0$; so, the dt_i being linearly independent, $\frac{\partial f}{\partial T_i}(t) = 0$ for $1 \leqslant i \leqslant r$; hence, $\deg(f)$ being minimal, $\frac{\partial f}{\partial T_i} = 0$ for $1 \leqslant i \leqslant r$. If k has characteristic 0, it follows that $f = 0$; hence, $t_1,\ldots,t_r$ are algebraically independent and $r \leqslant n$.

If k has characteristic $p > 0$, then $f = h(T_1^p,\dots,T_r^p)$. If $f(T) = \Sigma c_{(i)} T_1^{pi_1} \dots T_r^{pi_r}$, let $d_{(i)} = \sqrt[p]{c_{(i)}}$ and $g = \Sigma d_{(i)} \otimes t_1^{i_1} \dots t_r^{i_r} \in k' \otimes_k K$; then $g^p = 0$; so, since $k' \otimes_k K$ is reduced, $g = 0$. If $d_{(i)} = \Sigma e_{(i),j} f_j$ where the f_j are linearly independent over k, then $\Sigma e_{(i),j} t_1^{i_1} \dots t_r^{i_r} = 0$ for any j, contradicting the minimality of $\deg(f)$. Hence, $t_1,\dots,t_r$ are algebraically independent and $r \leqslant n$, completing the proof.

Corollary (6.6). - Let K be a finitely generated field extension of k and $n = \mathrm{tr.deg}_k K$. Then $\dim_K(\Omega^1_{K/k}) \geqslant n$, with equality if and only if K/k is separably generated.

Proof. If $\dim_K(\Omega^1_{K/k}) = r \leqslant n$, then $\Omega^1_{K/k}$ is a K-module with n generators and, by (6.5) is free of rank n. Thus $r = n$.

Corollary (6.7). - An algebraic k-scheme X is smooth if and only if $\Omega^1_{X/k}$ is locally free and the local rings of the generic points are separable field extensions of k.

Proof. The assertion results from (6.4), (6.5), and (III,2.8).

Chapter VIII - Curves

1. The Riemann-Roch theorem

Definition (1.1). - Let k be an artinian ring, X a proper k-scheme and F a coherent sheaf on X. The Euler-Poincaré characteristic of F, denoted $\chi(F)$, is defined as the alternating sum $\Sigma(-1)^i h^i(F)$ of the length $h^i(F)$ of the k-modules $H^i(X,F)$. If D is a divisor on X, then we often write $\chi(D)$ (resp. $h^i(D)$) in place of $\chi(O_X(D))$ (resp. $h^1(O_X(D))$).

Proposition (1.2). - Let k be an artinian ring, X a proper curve over k and $D_1,\ldots,D_r$ divisors on X. Then the Euler-Poincaré characteristic $\chi(n_1D_1 + \ldots + n_rD_r)$ is a linear polynomial in $n_1,\ldots,n_r$ with integer coefficients.

Proof. If $r = 0$, the assertion is trivial. If $r \geq 1$, let $J = O_X(-D_1)\cap O_X$, $J' = J(D_1)$, $F = O_X/J$ and $G = (O_X/J')(-D_1)$. Since the sequences

$$0\to J(n_1D_1+\ldots+n_rD_r)\to O_X(n_1D_1+\ldots+n_rD_r)\to F(n_1D_1+\ldots+n_rD_r)\to 0$$

$$0\to J'((n_1-1)D_1+\ldots+n_rD_r)\to O_X((n_1-1)D_1+\ldots+n_rD_r)\to G(n_1D_1+\ldots+n_rD_r)\to 0$$

are exact, and since $\dim(\mathrm{Supp}(F)) = \dim(\mathrm{Supp}(G)) = 0$,

$$\chi(n_1D_1+ \ldots + n_rD_r)-\chi((n_1-1)D_1+ \ldots + n_rD_r)$$

is a constant. Therefore, the assertion follows by induction.

Definition (1.3). - Let k be an artinian ring, X a proper curve over k and D a divisor on X. Then the degree of D is defined as the leading coefficient of the polynomial $\chi(nD)$.

Theorem (1.4) (Riemann). - Let k be an artinian ring, X a proper curve over k and D a divisor on X. Then

$$\chi(D) = \deg(D) + \chi(O_X).$$

Proposition (1.5). - Let k be an artinian ring, X a proper curve over k and C, D two divisors on X. Then, $\deg(C-D) = \deg(C)-\deg(D)$.

Proof. By taking successively $n = 0$ and $m = 0$ in the polynomial $\chi(mC-nD) = am-bn+c$, it follows that $a = \deg(C)$ and $b = \deg(D)$; by taking $m = n$, it follows that $a-b = \deg(C-D)$.

Proposition (1.6). - Let k be an artinian ring, X a proper normal curve over k and D a divisor on X. Then, $\deg(D) = \Sigma v_x(D)\deg_k(x)$ where $v_x(D)$ is the integer defined in (VII,3.9) and $\deg_k(x)$ is the k-length of $k(x)$.

Proof. By (VII,2.6), (VII,3.10,(iii)) and (1.5), we may assume $cyc(D) = x$. Since the sequence

$$0 \longrightarrow O_X \longrightarrow O_X(D) \longrightarrow k(x) \longrightarrow 0$$

is exact, $\chi(D)-\chi(O_X) = \deg_k(x)$; hence, by (1.4), $\deg(D) = \deg_k(x)$.

Remark (1.7). - Let X be a curve, F a subsheaf of K_X such that $F_{x_0} = K_{x_0}$ for all generic points x_0 of X and G the quotient K_X/F. Then there exists an injection $G \longrightarrow \Pi_{x \text{ closed}} G'_x$ where G'_x is the O_X-Module whose stalks are G_x at x and 0 elsewhere. Since there is an injection $\oplus G'_x \longrightarrow \Pi G'_x$ and since $\oplus G'_x$ and G have the same stalks, there exists a canonical isomorphism $G \xrightarrow{\sim} \oplus G'_x$.

Proposition (1.8). - Let X be an S_1 noetherian curve, $K = \Gamma(X,K_X)$ and F a subsheaf of K_X such that $F_{x_0} = K_{x_0}$ for all generic points x_0 of X. Then there exists an exact sequence

$$0 \longrightarrow H^0(X,F) \longrightarrow K \longrightarrow \oplus (K_x/F_x) \longrightarrow H^1(X,F) \longrightarrow 0.$$

Proof. The assertion results from the exact sequence $0 \longrightarrow F \longrightarrow K_X \longrightarrow K_X/F \longrightarrow 0$ because $H^0(X,K_X/F) = \oplus K_X/F_x$ by (1.7) and $H^1(X,K_X) = 0$ by (VII,3.8).

Remark (1.9). - Let k be an artinian ring, X an S_1 curve of finite type over k, $K = \Gamma(X,K_X)$ and F a coherent subsheaf of K_X. It follows from (1.8) applied to F that $H^1(X,F)^*$ may be identified with the set $J(F)$ of families δ of maps δ_x, one for each closed point x of X, which satisfy the following four conditions:

(a) $\delta_x : K \longrightarrow k$ is a k-linear map.

(b) $\delta_x(K_{x_0}) = 0$ for all generic points x_0 such that $x_0 \not\longrightarrow x$ or such that $x_0 \notin \mathrm{Supp}(F)$.

(c) $\delta_x(F_x) = 0$.

(d) $\sum_x \delta_x(f) = 0$ for each $f \in K$.

A family $\delta \in J(F)$, for some F, is called a pseudo-differential. The set J of all pseudo-differentials has a natural K-module structure: If $\delta \in J$ and $f \in K$, then $(f\delta)_x(g) = \delta_x(fg)$ for $g \in K$. It is easily seen that, if $\delta \in J(F)$, then $f\delta \in J(G)$ where $G_x = \{g \in O_x | fg \in F_x\}$.

If $F \subset F' \subset K_X$ and $\mathrm{Supp}(F) = \mathrm{Supp}(F')$, then, clearly, $J(F') \subset J(F)$. If $F' = F + \mathrm{ann}(F)$, then $J(F) \subset J(F')$ and $\mathrm{Supp}(F') = X$. If $\mathrm{Supp}(F) = X$, then, for each $x \in X$, F_x contains a non-zero-divisor f_x of K; moreover, since $F_x = O_x$ for almost all x, almost all f_x may be taken as 1. Then, the f_x^{-1} define a divisor D such that $O_X(D) \subset F$. Therefore, $J = \cup J(D)$ where $J(D) = J(O_X(D))$ and D runs through $\mathrm{Div}(X)$.

<u>Proposition (1.10)</u>. - Let k be an artinian ring, X an S_1 curve proper over k, $K = \Gamma(X,K_X)$ and J the K-module of pseudo-differentials. Then, $\mathrm{rank}_K(J) \leqslant 1$.

<u>Proof.</u> Suppose $\delta_1,\ldots,\delta_r \in J$ are linearly independent over K. Let D be a divisor such that $\delta_1,\ldots,\delta_r \in J(D)$. Then, for any divisor C, $J(D-C) \supset H^0(C)\delta_1 + \ldots + H^0(C)\delta_r$. Hence, $h^1(D-C) \geqslant rh^0(C)$. Replacing C by $D-C$ yields $h^1(C) \geqslant rh^0(D-C)$. Thus, by Riemann's theorem (1.4),

$$-[\deg(D-C) + \chi(O_X)] + h^0(D-C) \geqslant r[\deg(C) + \chi(O_X) + h^1(C)]$$

and so

(1.10.1) $$-\deg(D) \geqslant (r-1)\deg(C) + (r+1)\chi(O_X) + (r^2-1)h^0(D-C).$$

Now, if we let $\deg(C) \longrightarrow \infty$, we see that $r \leqslant 1$.

<u>Proposition (1.11)</u>. - Let k be a field, X a connected normal curve proper over k and δ a nonzero pseudo-differential. Then there exists a unique maximal divisor D such that $\delta \in J(D)$. This divisor is denoted (δ) and is called a <u>canonical divisor</u>. Moreover, $v_x((\delta))$ is the largest integer n such that $\delta_x(t_x^{-n}O_x) = 0$ where t_x is a uniformizing parameter at x.

<u>Proof</u>. With $r = 1$, (1.10.1) yields that, if there exists a $\delta \in J(D)$, then $\deg(D) \leqslant -2\chi(O_X)$. However, it is easily seen that if $\delta \in J(D)$ and $\delta \in J(D')$, then $\delta \in J(\mathrm{Max}(D,D'))$; whence, the assertion.

<u>Remark (1.12)</u>. - Let k be an artinian ring and X an S_1 curve of finite type over k. For each open set U of X, let $J_X(U)$ (resp. $\omega_X(U)$) be the set of pseudo-differentials δ on the scheme-theoretic closure of U (resp. such that $\delta_x(O_x) = 0$ for all closed points $x \in U$). It is easily seen that the $J_X(U)$ (resp. $\omega_X(U)$) form

a sheaf, called the <u>sheaf of rational pseudo-differentials</u> (resp. <u>sheaf of regular pseudo-differentials</u> or <u>canonical sheaf</u>).

<u>Proposition (1.13)</u>. - Let k be a field, X a connected normal algebraic curve over k and δ a nonzero pseudo-differential. Then the map $K \longrightarrow J$ defined by $f \longmapsto f\delta$ induces an isomorphism $O_X((\delta)) \xrightarrow{\sim} \omega_X$.

<u>Proof</u>. For any closed point $x \in X$, the following conditions are clearly equivalent: $f\delta \in \omega_x$; $(f\delta)_x(O_x) = 0$; $\delta_x(fO_x) = 0$; and $f \in (O_X(\delta))_x$. Surjectivity results from (1.10).

<u>Remark (1.14)</u>. - Let k be a field and X a reduced algebraic curve over k. It follows from (1.10) applied componentwise that we may identify J_x with K_x and J with K. So, by (1.8), there exists an exact sequence

$$0 \longrightarrow \Gamma(X,\omega_X) \longrightarrow J \longrightarrow \oplus J_x/\omega_x \longrightarrow H^1(X,\omega_x) \longrightarrow 0.$$

Now, for each closed point $x \in X$ and each $\delta \in J$, let $\mathrm{Res}_x(\delta) = \delta_x(1)$. Then $\mathrm{Res}_x : J \longrightarrow k$ is k linear and $\Sigma \mathrm{Res}_x(\delta) = \Sigma\delta_x(1) = 0$; furthermore, if $\delta_x \in \omega_x$, then $\mathrm{Res}_x(\delta) = \delta_x(1) = 0$. Hence, $\mathrm{Res} = (\mathrm{Res}_x) : H^1(X,\omega_X) \longrightarrow k$; Res is called the <u>residue map</u> of X.

<u>Theorem (1.15) (Roch)</u>. - Let k be a field, X a reduced curve proper over k and F a coherent subsheaf of K_X. Then the map

$$\Psi : H^0(X,\underline{Hom}(F,\omega_X)) \longrightarrow H^1(X,F)^*,$$

induced by Res, is an isomorphism.

<u>Proof</u>. Given $\delta \in H^1(X,F)^* = J(F)$, define $\varphi(\delta) : F \longrightarrow \omega_X$ by $\varphi(\delta)_x(f) = f\delta$ for all $x \in X$ and $f \in F_x \subset K_x \subset K$. Then, for any closed point $y \in X$, $\mathrm{Res}_y(\varphi(\delta)_y(f)) = (f\delta)_y(1) = \delta_y(f)$; so, $\Psi \circ \varphi =$

$= id_{H^1(X,F)^*}$. Finally, if $u : F \longrightarrow \omega_X$, then, for $f \in F_x$, $(\varphi(\Psi(u))_x(f))_x = (f\Psi(u))_x = u(f)_x$; so, $\varphi \circ \Psi = id_{Hom(F,\omega_X)}$.

Proposition (1.16) (Rosenlicht). - Let k be a field and X a reduced algebraic curve over k. Let Y be the normalization of X and $p : Y \longrightarrow X$ the canonical morphism. Then:

(i) The O_X-homomorphism $\varphi: p_*\omega_Y \longrightarrow \omega_X$, defined by $\varphi(\delta)_x = \sum_{p(y)=x} \delta_y$ is an injection.

(ii) The natural pairing $(p_*O_Y/O_X) \times (\omega_X/p_*\omega_Y) \longrightarrow k$ is nonsingular.

(iii) ω_X is coherent.

(iv) Let $C = Ann(p_*O_Y/O_X)$, $n_x = \dim_k(p_*O_Y/C)_x$ and $d_x = \dim_k(p_*O_Y/O_X)_x$. Then, for all singular points x of X, $d_x + 1 \leqslant n_x \leqslant 2d_x$ and the equality $n_x = 2d_x$ holds if and only if ω_X is free of rank 1 at x.

Proof. Let x be a closed point of X, $A = (p_*O_Y)_x$, q the radical of A, $\{y_1,\dots,y_n\} = p^{-1}(x)$ and $A_i = O_{y_i}$. For any integer $r>0, A/q^r = \Pi A_i/q^r A_i$ by (II,4.9); hence, given $g \in A_i$, there exists $h \in A$ such that $h \equiv g \bmod q^r A_i$ and $h \equiv 0 \bmod q^r A_j$ for $j \neq i$. If δ is a pseudo-differential on Y, there is an r such that $\delta_{y_i}(q^r A_i) = 0$ for all i. Therefore, if I is an ideal of A which contains q^r for some r, then $\varphi(\delta)_x(I) = 0$ implies $\delta_{y_i}(IA_i) = 0$ for all i.

Generically, φ is a map of one-dimensional vector spaces by (1.10). It now follows that φ is injective and that any pseudo-differential α on X is the form $\varphi(\delta)$ for some pseudo-differential δ on Y. Moreover, if $\alpha_x(A) = 0$, then $\delta_{y_i}(A_i) = 0$ for all i; so, $\alpha_x \in (p_*\omega_Y)_x$. Therefore, if B is any k-subspace of A and $\omega(B)$ is the set of pseudo-differentials α on X such that $\alpha_x(B) = 0$, then

$$(p_*\omega_Y)_x = \omega(A)$$

and the natural pairing gives rise to the injection

$$\omega(B)/\omega(A) \longrightarrow (A/B)^*.$$

Since $\omega(O_x) = \omega_x$ and $C_x \subset O_x$, it follows that to prove (ii) it suffices to prove that $\dim_k(\omega(C_x)/\omega(A)) = \dim_k(A/C_x)$. However, $\alpha = \varphi(\delta) \in \omega(C_x)$ if and only if $\delta_{y_i}(C_xA_i) = 0$ for all i. Since by (VII,2.6), C_xA_i is principal and $\mathrm{rank}_{K_{y_i}}(J_{y_i}) = 1$, $\dim(\{\delta \in J_{y_i} | \delta_{y_i}(C_xA_i) = 0\}/\omega_{y_i}) = \dim(A_i/C_xA_i)$; whence (ii).

Assertion (iii) results immediately from (i) and (ii).

To prove (iv), note that, if x is singular, then $k + C_x \subset O_x \subset A$; whence, $d_x + 1 \leq n_x$. For each i, let $\delta_i \in \omega_x$ generate $A_i\omega_x$. Making a purely transcendental extension of the ground field, if necessary, we may assume it is infinite; then, a suitable combination δ of the δ_i generates all $A_i\omega_x$ and $A\delta = A\omega_x$. Let $f \in A$ and suppose $f\delta \in \omega(A)$. Then $(A\delta)(f) = 0$, so $f \in O_x$ by (ii). However, $C_x = \mathrm{ann}(\omega_x/\omega(A))$ by (ii). Therefore, $f \in C_x$.

The map $f \mapsto f\delta$ defines an injection $u : O_x/C_x \longrightarrow \omega_x/\omega(A)$; hence, $n_x - d_x \leq d_x$. If ω_x is a free O_x-module of rank 1, then necessarily δ is a basis; so u is surjective and $n_x - d_x = d_x$. Conversely, if this equality holds, then u is surjective and every $\alpha \in \omega_x$ is of the form $g\delta + \beta$ where $g \in O_x$ and $\beta \in \omega(A)$. However, $\beta = f\delta$ for some $f \in A$; so, $f \in O_x$ and $\alpha = (g + f)\delta$.

<u>Remark (1.17)</u> - (i) Under the conditions of (1.16), suppose X is integral and proper and let $\pi = h^1(O_X)$ (resp. $g = h^1(O_Y)$) be the arithmetic (resp. geometric) genus of X. Then the exact sequence $0 \longrightarrow O_X \longrightarrow p_*O_Y \longrightarrow p_*O_Y/O_X \longrightarrow 0$ shows that

$$\pi = g + \Sigma d_x.$$

(ii) Let X be a reduced algebraic curve lying on a smooth algebraic scheme P of pure dimension r. Then it follows from (1.16), (1.15) and (I,2.1;2.3; and 4.6) that the sheaf ω_X of regular pseudo-differentials is of the form $\underline{\mathrm{Ext}}^{r-1}_{O_P}(O_X,\Omega^r_{P/k})$. Moreover, by (I,2.6),(III,4.5 and 4.12), and (VII,6.2) ω_X is locally free of rank 1 at $x \in X$ (or, equivalently, $n_x = 2d_x$) if X is a complete intersection in P locally at x.

In particular, ω_X is invertible if X is a complete intersection in $\mathbb{P}^r$ (Rosenlicht) or if X lies on a smooth surface F (Gorenstein-Samuel); further, if K_F is a canonical divisor on F (i.e., $\Omega^2_{F/k} = O_F(K_F)$), then $X.(X + K_F)$ is a canonical divisor on X (I,2.4).

2. Tate's definition of residues

Remark (2.1). - Let k be a ring, A a k-algebra and M, N two A-modules. Then there is a natural left (resp. right) A-module structure on $\mathrm{Hom}_k(M,N)$: If $u \in \mathrm{Hom}_k(M,N)$, $a \in A$ and $x \in M$, then $(au)(x) = au(x)$ (resp. $(ua)(x) = u(ax)$). Let $[A,\mathrm{Hom}_k(M,N)]$ denote the k-submodule of $\mathrm{Hom}_k(M,N)$ generated by all elements of the form au-ua.

Proposition (2.2). - Let k be a ring, A a k-algebra and $0 \longrightarrow N \xrightarrow{j} E \xrightarrow{p} M \longrightarrow 0$ an exact sequence of A-modules. If M is k-projective, then there exists a canonical A-homomorphism

$$\varphi : \Omega^1_{A/k} \longrightarrow H = \mathrm{Hom}_k(M,N)/[A,\mathrm{Hom}_k(M,N)]$$

such that $\varphi(dt) = j^{-1}\circ(t\sigma-\sigma t)$ for any k-section σ of p.

Proof. Define $D_\sigma : A \longrightarrow H$ by $D_\sigma(t) = j^{-1}\circ(t\sigma-\sigma t)$; D_σ is well-defined because $p\circ(t\sigma-\sigma t) = t-t = 0$. If σ' is another k-section, let $\tau = \sigma' - \sigma$. Then $p \circ \tau = 0$; so, $\varrho = j^{-1} \circ \tau \in \mathrm{Hom}_k(M,N)$. Now, $D_{\sigma'}(t) - D_{\sigma}(t) = j^{-1} \circ (t\sigma' - \sigma' t - t\sigma + \sigma t) = t\varrho - \varrho t \in [A, \mathrm{Hom}_k(M,N)]$. Thus $D = D_\sigma$ is independent of σ. If $t, t' \in A$, then $D(tt') = j^{-1}(tt'\sigma - t\sigma t' + t\sigma t' - \sigma tt') = tD(t') + t'D(t)$. Thus, D is a k-derivation; whence, the assertion.

<u>Definition (2.3)</u>. - Let k be a ring and A a k-algebra. Then define S_A as the set of all $s \in A$ satisfying the following two conditions:

(a) s is a non-zero-divisor.

(b) A/sA is projective of finite rank over k.

<u>Lemma (2.4)</u>. - Let k be a ring and A a k-algebra. Then S_A is a multiplicative set.

<u>Proof</u>. Let $r, s \in S_A$. Then, clearly, rs is a non-zero-divisor. Furthermore, the sequence

(2.4.1) $$0 \longrightarrow A/sA \xrightarrow{r} A/rsA \longrightarrow A/rA \longrightarrow 0$$

is exact; hence, A/rsA is k-projective of finite rank.

<u>Definition (2.5)</u>. - Let k be a ring, A a k-algebra, $\omega \in \Omega^1_{A/k}$ and $s \in S_A$. Then $\mathrm{Res}_{A/k}(\omega/s)$ is defined as $\mathrm{tr}_{(A/sA)/k}(\varphi(\omega))$ where φ is defined as in (2.2) with respect to $0 \longrightarrow A/sA \xrightarrow{s} A/s^2A \longrightarrow A/sA \longrightarrow 0$.

<u>Remark (2.6)</u>. - Let k be a ring and A a k-algebra. Then

(i) $\mathrm{Res}_{A/k}(\omega/1) = 0$ for any $\omega \in \Omega^1_{A/k}$.

(ii) $\mathrm{Res}_{A/k}(adt/s) = \mathrm{tr}_{(A/sA)/k}(s^{-1}(t\sigma - \sigma t)a)$ where $a, t \in A$, $s \in S_A$ and σ is a k-section of $A/s^2A \longrightarrow A/sA$.

<u>Lemma (2.7)</u>. - Let k be a ring, A a k-algebra and $u : A \longrightarrow A$ a k-linear map. Suppose $u(rA) \cap rA = 0$ and $u(sA) \cap sA = 0$

for $r, s \in S_A$. Then $tr_{(A/rA)/k}(u_r) = tr_{(A/sA)/k}(u_s)$ where $u_t = u \otimes id_{(A/tA)}$.

<u>Proof</u>. By symmetry, we may replace s by rs. Then, since (2.4.1) splits, the corresponding matrix $M(u_s)$ has the form $\begin{pmatrix} u_r & * \\ * & 0 \end{pmatrix}$; whence, the assertion.

<u>Definition (2.8)</u>. - Let k be a ring. A a k-algebra and $u : A \longrightarrow A$ a k-linear map such that $u(sA) \cap sA = 0$ for some $s \in S_A$. Then the <u>trace</u> of u, denoted $tr_{A/k}(u)$, is defined as the element $tr_{(A/sA)/k}(u_s) \in k$ where $u_s = u \otimes id_{A/sA}$.

<u>Proposition (2.9)</u>. - Let k be a ring, A a k-algebra and $\Pi : A \longrightarrow sA$ a k-linear projection. Then for all $a, t \in A$,

$$Res_{A/k}(adt/s) = tr_{A/k}(s^{-1}(\Pi t - t\Pi)a).$$

<u>Proof</u>. If $R = \ker(\Pi)$, then $A = R \oplus sR \oplus s^2A$, $R \cong A/sA$ and $R \oplus sR \cong A/s^2A$; whence $\sigma' = id_A - \Pi$ induces a k-section σ of $A/s^2A \longrightarrow A/sA$. Since $\Pi t - t\Pi = t\sigma' - \sigma' t$, it follows that $tr_{A/k}(s^{-1}(\Pi t - t\Pi)a) = tr_{(A/sA)/k}(s^{-1}(t\sigma - \sigma t)a) = Res_{A/k}(adt/s)$.

<u>Proposition (2.10)</u>. - Let k be a ring and A a k-algebra and $K = S_A^{-1}A$. Then $Res_{A/k}$ is a k-linear map from $\Omega^1_{K/k}$ to k.

<u>Proof</u>. Let $a, t \in A$ and $r, s \in S_A$. Let Π be a k-linear projection $A \longrightarrow rsA$. Then $r^{-1}\Pi r$ is a k-linear projection $A \longrightarrow sA$. Hence, by (2.9), $Res_{A/k}(radt/rs) = tr_{A/k}(s^{-1}r^{-1}(\Pi t - t\Pi)ra)$ and $Res_{A/k}(adt/s) = tr_{A/k}(s^{-1}((r^{-1}\Pi r)t - t(r^{-1}\Pi r))a)$. Therefore, by (2.7), $Res_{A/k}(radt/rs) = Res_{A/k}(adt/s)$; whence, the assertion.

<u>Proposition (2.11)</u>. - Let k be a ring and A a k-algebra. Then $Res_{A/k}(ads/s) = tr_{(A/sA)/k}(a)$ for all $a \in A$, $s \in S_A$. In particular, $Res_{A/k}(ds/s) = rank_k(A/sA).1_k$.

Proof. Let $\Pi : A \longrightarrow sA$ be a k-linear projection and $\sigma' = id_A - \Pi$. Then $\Pi t - t\Pi = t\sigma' - \sigma' t$, $\sigma' \circ s = 0$, $tr_{A/k}(s^{-1}(s\sigma' - \sigma' s)a) = tr_{A/k}(\sigma' a)$ and $M(\sigma' a) = \begin{pmatrix} a & * \\ * & 0 \end{pmatrix}$; whence, the assertion.

Proposition (2.12). - Let k be a ring, A a k-algebra and $s \in S_A$. Then $Res_{A/k}(ds/s^n) = 0$ for $n > 1$.

Proof. Decompose A into a k-direct sum $A = T \oplus s^{n-1}R \oplus s^n A$ where $T = R \oplus sR \oplus \ldots \oplus s^{n-2}R$ and let Π be the projection $A \longrightarrow s^n A$. If $a = t + s^{n-1}r + s^n b$ is the decomposition of $a \in A$, then $u(a) = s^{-n}(\Pi s - s\Pi)a = s^{-n}(s^n r + s^{n+1}b - s^{n+1}b) = r$; hence,

$$M(u) = \begin{pmatrix} 0 & s^{1-n} & 0 \\ 0 & 0 & 0 \\ 0 & 0 & 0 \end{pmatrix} \quad \text{and} \quad tr_{A/k}(u) = 0.$$

3. Functorial properties of residues

Lemma (3.1). - Let k be a ring and $\varphi : A \longrightarrow A'$ a k-algebra homomorphism. Let $s \in S_A$ and $\omega \in \Omega^1_{A/k}$, let $s' = \varphi(s)$ and $\omega' = \varphi(\omega)$. Assume:

(a) s' is a non-zero-divisor in A'

(b) φ induces an isomorphism $A/s^2A \xrightarrow{\sim} A'/s'^2A'$.

Then $s' \in S_{A'}$ and $Res_{A'/k}(\omega'/s') = Res_{A/k}(\omega/s)$.

Proof. In the commutative diagram induced by φ,

$$\begin{array}{ccccccccc} 0 & \longrightarrow & A/sA & \longrightarrow & A/s^2A & \longrightarrow & A/sA & \longrightarrow & 0 \\ & & \downarrow & & \downarrow & & \downarrow & & \\ 0 & \longrightarrow & A'/s'A' & \longrightarrow & A'/s'^2A' & \longrightarrow & A'/s'A' & \longrightarrow & 0 \end{array} ,$$

the vertical maps are isomorphisms; whence, the assertion.

<u>Proposition (3.2)</u>. - Let k be a ring, A a k-algebra, $s \in S_A$ and Q a multiplicative set in A such that $\mathrm{Spec}(A/sA) \subset \mathrm{Spec}(Q^{-1}A)$. Then $\mathrm{Res}_{A/k}(\omega/s) = \mathrm{Res}_{Q^{-1}A/k}(\omega/s)$ for all $\omega \in \Omega^1_{A/k}$.

<u>Proof</u>. Since localization is exact, s is a non-zero-divisor in $Q^{-1}A$ and $Q^{-1}A/s^2Q^{-1}A = Q^{-1}(A/s^2A) = A/s^2A$; whence, the assertion results from (3.1).

<u>Proposition (3.3)</u>. - Let k be a ring and A a noetherian k-algebra. Let $s \in S_A$ and m an ideal contained in sA. If $\hat{A} = \varprojlim(A/m^r)$, then $\mathrm{Res}_{\hat{A}/k}(\omega/s) = \mathrm{Res}_{A/k}(\omega/s)$ for all $\omega \in \Omega^1_{A/k}$.

<u>Proof</u>. By (II,1.17), (3.1a) holds and by (II,1.19), (3.1b) holds; whence, the assertion.

<u>Proposition (3.4)</u>. - Let k be a ring, $\{A_i\}$ a finite family of k-algebras and $A = \Pi A_i$. If $s = \Pi s_i$ where $s \in S_A$ and $s_i \in A_i$ and if $\omega = \Sigma\omega_i$ where $\omega_i \in \Omega^1_{A_i/k}$, then $s_i \in S_{A_i}$ and $\mathrm{Res}_{A/k}(\omega/s) = \Sigma \mathrm{Res}_{A_i/k}(\omega_i/s_i)$.

<u>Proof</u>. Since $A/sA = \Pi A_i/s_iA_i$, it follows that $s \in S_A$ (if and) only if $s_i \in S_{A_i}$ for each i. Choose splittings σ_i of $A_i/(s_i)^2A_i \longrightarrow A_i/s_iA_i$; then $\sigma = \Pi\sigma_i$ is a splitting of $A/s^2A \longrightarrow A/sA$. By linearity of Res, we may assume $\omega = adt$. Let $a = \Pi a_i$ and $t = \Pi t_i$ where a_i, $t_i \in A_i$. Then $\mathrm{Res}_{A/k}(\omega/s) =$
$= \mathrm{tr}_{A/k}(\Pi s_i^{-1}(\sigma_i t_i - t_i\sigma_i)a_i) = \Sigma \mathrm{Res}_{A_i/k}(\omega_i/s_i)$.

<u>Proposition (3.5)</u>. - Let k be a ring, A a noetherian k-algebra of dimension 1 and $X = \mathrm{Spec}(A)$. If $\omega \in \Omega^1_{A/k}$ and $s \in S_A$, then $\mathrm{Res}_{A/k}(\omega/s) = \sum_{x \text{ closed}} \mathrm{Res}_x(\omega/s)$ where $\mathrm{Res}_x(\omega/s) = \mathrm{Res}_{O_x/k}(\omega/s)$.

<u>Proof</u>. The sum is finite because, by (2.6(i)) and (2.10), whenever $s(x) \neq 0$, $\mathrm{Res}_x(\omega/s) = 0$. Let $\{x_i\}$ be the zeros of s, $m = sA$ and

$\hat{A} = \varprojlim(A/m^r)$. Then, by (VI,6.7) and (II,1.24), $\hat{A} = \Pi\hat{O}_{x_i}$. Therefore, the assertion results from (3.3) and (3.4).

Proposition (3.6). - Let k be a ring, A, k' two k-algebras and $A' = A\otimes_k k'$. If $s \in S_A$ and $\omega \in \Omega^1_{A/k}$, then $s' = s\otimes 1 \in S_{A'}$ and $\mathrm{Res}_{A/k}(\omega/s)\otimes 1 = \mathrm{Res}_{A'/k'}(\omega\otimes 1/s\otimes 1)$.

Proof. Since $A'/s'A' = (A/sA)\otimes_k k'$, $A'/s'A'$ is k' projective of finite rank. Further, the exact sequence $0\longrightarrow A \xrightarrow{s} A \longrightarrow A/sA \longrightarrow 0$ is k-split; so, the sequence $0\longrightarrow A' \xrightarrow{s'} A'\longrightarrow A'/s'A' \longrightarrow 0$, obtained by tensoring it with k', is exact.

Choose a k-splitting σ of $A/s^2A\longrightarrow A/sA$; then $\sigma' = \sigma\otimes 1$ is a k-splitting of $A'/(s')^2A'\longrightarrow A'/s'A'$. We may assume $\omega = adt$. Then by (VI,6.5) $\mathrm{Res}_{A'/k}(\omega\otimes 1/s\otimes 1) = \mathrm{tr}_{A'/k'}(s^{-1}(\sigma t-t\sigma)a\otimes 1) =$
$= \mathrm{tr}_{A/k}(s^{-1}(\sigma t-t\sigma)a)\otimes 1 = \mathrm{Res}_{A/k}(\omega/s)\otimes 1$.

Proposition (3.7) (The trace formula). - Let k be a ring and $\varphi : A\longrightarrow A'$ a homomorphism of k-algebras. Suppose A' is projective of finite rank over A. Let $\mathrm{Tr}_{A'/A}$ be the homomorphism $\mathrm{id}_{\Omega^1_{A/k}}\otimes\mathrm{tr}_{A'/A} : \Omega^1_{A/k}\otimes_A A'\longrightarrow\Omega^1_{A/k}$. If $\omega \in \Omega^1_{A/k}\otimes_A A'$, $s \in S_A$, and $s' = \varphi(s)$, then $s' \in S_A'$ and

$$\mathrm{Res}_{A'/k}(\omega/s') = \mathrm{Res}_{A/k}(\mathrm{Tr}_{A'/A}(\omega)/s).$$

Proof. Clearly, s' is a non-zero-divisor in A'. Since A' is a direct summand of A^p, $A'/s'A'$ is a direct summand of $(A/sA)^p$; hence, $A'/s'A'$ is k-projective of finite rank.

Let $\Pi : A\longrightarrow sA$ be a k-linear projection . Then $\Pi' = \Pi\otimes\mathrm{id}_{A'}: A'\longrightarrow s'A'$ is a k-linear projection. Since $\mathrm{Tr}_{A'/A}$ is linear, we may assume $\omega = a'dt$ where a', $t \in A$. Let

$\varphi = s^{-1}(\Pi t - t\Pi)$. Since $tr_{A'/A}$ is A-linear, $tr_{A'/k}((\varphi\otimes id_{A'})a') =$ $= tr_{A/k}(tr_{A'/A}((\varphi\otimes id_{A'})a)) = tr_{A/k}(\varphi(tr_{A'/A}(a')))$. Therefore, by (2.9), $Res_{A'/k}(a'dt/s') = Res_{A/k}(tr_{A'/A}(a')dt/s) =$ $= Res_{A/k}(Tr_{A'/A}(a'dt)/s)$.

4. Residues on algebraic curves

Example (4.1). - Let k be a field, T an indeterminate, $P(T)$ a monic irreducible polynomial and $d = \deg(P)$. Let $r(T)/s(T) =$ $= (r_m(T)/P(T)^m) + \ldots + (r_1(T)/P(T)) + (r_0(T)/s_0(T))$ be a rational function such that $\deg(r_i(T)) < d$ for $i > 0$ and $s_0(T) \not\equiv 0 \bmod P(T)$ If $r_1(T) = a_{d-1}T^{d-1} + \ldots + a_0$, then $Res_x(rdP/s) = a_{d-1}$ where $x \in \mathbb{A}^1_k = Spec(k[T])$ is the closed point "cut out" by P.

Proof. By (2.6), $Res_x(r_0dP/s_0) = 0$; so, by (2.10), we may assume $r_0 = 0$. Let $P = \prod(T-b_j)^{q_j}$ where b_j are the distinct roots of P in a splitting field. Now, $r_i(T)/P(T)^i = \Sigma h_{ji}(T)$ where $h_{ji}(T) =$ $= (c_{ji}(T-b_j)^{q_j-1} + \ldots)/(T-b_j)^{q_j^i}$. Then, (3.6), (3.5), (2.10), (2.11) and (2.12), $Res_x(rdT/s) = \Sigma Res_{b_j}(c_{ji}d(T-b_j)/(T-b_j)^{q_j-1}) = \Sigma c_{ji}$; whence the assertion.

Proposition (4.2). - Let k be a field, X, Y two S_1 algebraic curves over k, $f : X \longrightarrow Y$ a covering map, $K = \Gamma(X,K_X)$, $L = \Gamma(Y,K_Y)$. Suppose f is flat (e.g., X integral and Y normal) and generically unramified. Then, for all $\omega \in \Omega^1_{K/k}$,

$$\Sigma Res_x(\omega) = \Sigma Res_y(Tr_{K/L}(\omega)).$$

Proof. We may assume $Y = Spec(O_y)$ and $X = Spec(A)$. Since f is generically étale, by (VI,4.9), $\Omega^1_{K/k} = \Omega^1_{L/k}\otimes_L K$. Furthermore,

$S = S_{O_y}$ is clearly the set of non-zero-divisors; so, $\Omega^1_{L/k} = S^{-1}\Omega^1_{O_y/k}$ and $K = S^{-1}A$. Therefore, the assertion follows from (3.7) and (3.5).

Theorem (4.3) (Residue formula). - Let k be a field, X a connected normal curve, proper over k and K its function field. Suppose K is separably generated over k. If $\omega \in \Omega^1_{K/k}$, then

$$\sum_{x \text{ closed}} \mathrm{Res}_x(\omega) = 0.$$

Proof. It follows from the hypothesis that there is a finite separable morphism $f : X \longrightarrow \mathbb{P}^1_k$. Therefore, by (4.2), we may assume $X = \mathbb{P}^1_k$. Further, by (3.6) and (3.5), we may assume k is algebraically closed.

Suppose $\omega = a\,dt$ where $a \in k(t)$. By decomposing a into partial fractions, using the linearity of Res and changing variables, we may assume that $a = t^n$, $n \geq 0$. However, t^n may have a pole only at ∞; so, by (2.6), $\mathrm{Res}_x(\omega) = 0$ for $x \neq \infty$. If $u = 1/t^n$, then $\omega = -du/u^{n+2}$; so, by (2.12), $\mathrm{Res}_\infty(\omega) = 0$.

Theorem (4.4). - Let k be a field and X a connected curve smooth and proper over k. Then $\Omega^1_{X/k} = \omega_X$ and the residue maps coincide.

Proof. Let $K = \Gamma(X, K_X)$ and $\omega \in \Omega^1_{K/k}$. For each $f \in K$ and $x \in X$ closed, let $\delta_x(f) = \mathrm{Res}_x(f\omega)$. Then, by (2.10), $\delta_x : K \longrightarrow k$ is a k-linear map and, by (4.3), $\Sigma\delta_x(f) = 0$ for all $f \in K$.

Let x be a closed point. Since X/k is smooth, $k(x)/k$ is separable. So, there exists $a \in k(x)$ such that $\mathrm{tr}_{k(x)/k}(a) \neq 0$. Let $b \in O_x$ have residue class a. If $\omega = (u/t_x^{n_x})dt_x$ where t_x is a uniformizing parameter of O_x and $u \in O_x^*$, then, by (2.11),

$\mathrm{Res}_x(f\omega) \neq 0$ for $f = bt_x^{n_x-1}u^{-1}$. Therefore, by (2.6), $\omega \in \Omega^1_{O_x/k} = O_x dt_x$ if and only if $\delta_x(O_x) = 0$.

Since $\omega \in \Omega^1_{O_x/k}$ for almost all x, the elements $t_x^{m_x}$, where $m_x = \max(0,n_x)$, define a divisor D such that $\delta = (\delta_x) \in J(-D)$. Therefore, since $\dim_K(J) = 1$ and $\dim_K(\Omega^1_{K/k}) = 1$, the map $\omega \longmapsto \varphi(\omega) = \delta$ defines an isomorphism $\Omega^1_{X/k} \xrightarrow{\sim} \omega_X$. Finally, $\mathrm{Res}_x(\varphi(\omega)) = \varphi(\omega)_x(1) = \mathrm{Res}_x(\omega)$.

Notation

$K_*(\underline{x})$, $K_*(\underline{x};M)$, $K^*(\underline{x};M)$, $H^*(\underline{x};M)$ (M an A-module, $x_i \in A$): I,4.

$gr^*(M)$, $gr_q^*(M)$ (M a filtered A-module, q an ideal): II,1.4.

$\varprojlim M_i$ ((M_i, f_j^i) a projective system): II,1.6.

$\hat{N}$ (N a filtered module): II,1.7.

rad(A) (A a ring): II,1.20.

Supp(F), Supp(M) (F a sheaf, M a module): II,2.1.

V(J) (J a sheaf of ideals): II,2.5.

Ass(M), Ass(F) (M a module, F a Module): II,3.1.

Ann(x): II,3.1.

$S^{-1}M$, $S^{-1}p$ (M an A-module, p a prime, $S \subset A$): II,3.9.

Q(p) (p a prime ideal): II,3.14.

$\ell_A(M)$, $\ell(M)$ (M an A-module): II,4.1.

$\chi(M,n)$: II,4.10.

$\Delta\chi$ (χ a polynomial) : II,4.11.

Q(M,n): II,4.11.

$P_{(M_n)}$: II,4.13.

$P_q(M,n)$: II,4.14.

dim(X), $\dim_A(M)$, dim(M) (X a topological space, M an A-module): III,1.1.

d(M), s(M): III,1.1.

$\mathrm{tr.deg}_k A$ (k a field, A a k-algebra): III,2.6.

$\mathrm{depth}_I(M)$, $\mathrm{depth}_A(M)$, depth(M) (M an A-module, I an ideal): IV,3.9, 3.11.

$\mathrm{proj.dim}_A(M)$, $\mathrm{inj.dim}_A(M)$ (M an A-module): III,5.1.

gl.hd(A) (A a ring): III,5.3.

$E^{\vee}$ (E a locally free sheaf): IV,2.6.

$y_r(F)$ (F a Module): IV,4.2.

ε^* : IV,5.2.

Terminology

BIBLIOGRAPHY

1. N. Bourbaki, "Algèbre Commutative", Hermann, Paris.
2. H. Cartan and S. Eilenberg, "Homological Algebra", Princeton University Press (1956).
3. P. Cartier, "Les groupes $Ext^s(A,B)$", Séminaire A. Grothendieck, No. 3, Secr. Math. I.H.P. Paris (1957).
4. R. Godement, "Topologie algébrique et théorie des faisceaux", Hermann, Paris (1958).
5. A. Grothendieck, "Sur quelque points d'algèbre homologique", Tohoku Math. J. IX (1957), 119-221.
6. ________, "Théorèmes de dualité pour les faisceaux algébriques cohérents", Séminaire Bourbaki, No. 149, Secr. Math I.H.P. Paris (1957).
7. ________, "Eléments de Géométrie Algébrique", (rédigés avec la collaboration de J. Dieudonné), Publ. Math. I.H.E.S., Paris (1960ff).
8. ________, "Local Cohomology", (mimeographed seminar notes by R. Hartshorne), Harvard (1961).
9. ________, "Séminaire de Géométrie Algébrique" (Notes polycopiés, prises par un groupe d'auditeurs), I.H.E.S. Paris (1960-1961).
10. R. Hartshorne, "Residues and Duality", Lectures Notes in Math., No. 20, Springer-Verlag (1966).
11. J.-P. Serre, "Groupes algébriques et corps de classes", Hermann, Paris (1959).
12. ________, "Algèbre Locale Multiplicités", (rédigé par P. Gabriel, Seconde édition), Lecture Notes in Math., No. 11, Springer-Verlag (1965).
13. J. Tate, "Algebraic Functions", (unpublished notes for Math 225), Harvard (1965).

Offsetdruck: Julius Beltz, Weinheim/Bergstr